Praise for *The Mind of the Market*

"[A] captivating raconteur of all the greatest hits of behavioral, evolutionary and neuropsychology, [and] provider of wonderful cocktail party material . . . Fascinating."

—*Los Angeles Times Book Review*

"Compelling . . . Take[s] us on an intimate tour of the best of the last half-century's work in behavioral economics and neuroscience."

—*New York Post*

"The book has no end of conversation starters, from capitalism as modern Darwinism to neuroeconomics that show that—biochemically, at least—a human brain is shockingly similar during smooth business deals and sex."

—*The Boston Globe*

"Extremely interesting . . . Shermer is a fantastic presenter."

—Steven D. Levitt, *The New York Times Freakonomics Blog*

"Michael Shermer brilliantly shows that the real experts of *Homo economicus* are often found in psychology, biology, even primatology."

—Frans de Waal, author of *Our Inner Ape*

"Have you ever wondered how people develop trust and live together peacefully? Michael Shermer's new book uses psychology and evolution to examine the root of these human achievements. . . . [He] has earned the right to our attention."

—*The Washington Post Book World*

"Drawing from research, and injecting his own wit, Shermer explains why people make bad decisions about money, why wealth can't buy you happiness, and why we love cooperating."

—*Psychology Today*

"Written with his customary verve and flair, *The Mind of the Market* is Michael Shermer at his best."

—Dinesh D'Souza, author of *What's So Great About America*

"[Shermer] does a bang-up job knitting together the complexities of science and the frail psychology of human beings to explain the unpredictable postmodern world of trade and finance. . . . An informative, inventive, broad-spectrum analysis of what makes modern man tick, starting with his wallet."

—*Kirkus Reviews*

"Economists who understand Charles Darwin are almost as rare as biologists who understand Adam Smith. Yet the two were essentially saying the same thing—that order emerges unordained from competition and innovation. Michael Shermer brilliantly brings the two insights together to explain how the human mind creates the human market."

—Matt Ridley, author of *The Origins of Virtue*

"Using fascinating examples . . . Shermer explores the evolutionary roots of our sense of fairness and justice, and shows how this rationale extends to the market. . . . Offers much insight into human behavior and rationales regarding money."

—*Publishers Weekly*

"The Ripley's Believe It or Not of behavioral economics, or why people act the way they do in a capitalistic marketplace . . . Shermer applies his wide-ranging knowledge of science and its rigorous investigatory discipline to uncover the answers and make connections between trade and emotion—in essence, popularizing neuron-economics."

—*Booklist*

THE MIND OF THE MARKET

THE MIND OF
THE MARKET

How Biology
and Psychology
Shape Our
Economic Lives

✦

MICHAEL SHERMER

A HOLT PAPERBACK

HENRY HOLT AND COMPANY ✦ NEW YORK

Holt Paperbacks
Henry Holt and Company, LLC
Publishers since 1866
175 Fifth Avenue
New York, New York 10010
www.henryholt.com

A Holt Paperback® and 🄯® are registered trademarks
of Henry Holt and Company, LLC.

For further information on the Skeptics Society and *Skeptic* magazine,
contact P.O. Box 338, Altadena, CA 91001; 626-794-3119;
www.michaelshermer.com; e-mail: skepticmag@aol.com; www.skeptic.com

Library of Congress Cataloging-in-Publication Data

Shermer, Michael.
The mind of the market : how biology and psychology shape our
economic lives / Michael Shermer.
      p.   cm.
Includes bibliographical references and index.
ISBN-13: 978-0-8050-8916-5
ISBN-10: 0-8050-8916-0
1. Economics—Psychological aspects.   2. Consumer behavior.   I. Title.
HB74.P8S46 2008
330.01'9—dc22   2008032782

Henry Holt books are available for special promotions and premiums.
For details contact: Director, Special Markets.

Originally published in hardcover in 2008 by Times Books

First Holt Paperbacks Edition 2009

Designed by Victoria Hartman

Printed in the United States of America
1 3 5 7 9 10 8 6 4 2

To

Jay Stuart Snelson

Aperire Terram Gentibus

A number of porcupines huddled together for warmth on a cold day in winter; but, as they began to prick one another with their quills, they were obliged to disperse. . . . At last, after many turns of huddling and dispersing, they discovered that they would be best off by remaining at a little distance from one another. In the same way the need of society drives the human porcupines together, only to be mutually repelled by the many prickly and disagreeable qualities of their nature. The moderate distance which they at last discover to be the only tolerable condition of intercourse is the code of politeness and fine manners.

—ARTHUR SCHOPENHAUER,
Parerga and Paralipomena, II, 31, 1851

CONTENTS

ECONOMICS FOR EVERYONE

In Jesus' parable of the talents, recounted in Matthew 25:14–29, the gospel author recalls the messiah as saying in the final verse, "For to everyone who has, more shall be given, and he will have an abundance; but from the one who does not have, even what he does have shall be taken away." Out of context this hardly sounds like the wisdom of the prophet who proclaimed that the meek shall inherit the earth, but in context, Jesus' point was that properly investing one's money (as measured in "talents") generates even more wealth. The servant who was given five talents invested it and gave his master ten talents in return. The servant who was given two talents invested it and gave his master four talents. But the servant who was given one talent buried it in the ground and gave his master back just the one talent. The master then ordered his risk-averse servant to give the one talent to the servant who had doubled his investment of five talents, and so he who earned the most was rewarded with even more. And thus it is that the rich get richer.

Jesus probably had in mind something more than an economic allegory about selecting the right investment vehicle for your money, but I want to employ the story as a parable about the mind of the market. In the 1960s, the sociologist of science Robert K. Merton conducted an extensive study of how scientific ideas are discovered and credited in the marketplace of ideas—in this case treating science as a market—and found that eminent scientists typically receive more credit than they deserve simply by dint of having a big name, while their junior colleagues and graduate students, who usually do most of the work, go largely unnoticed.[1] A similar well-known effect can be seen in how both innovative ideas and clever quotes gravitate up and are given credit to the most famous person associated with them.[2]

Merton called this the *Matthew Effect*. Marketers know it as *Cumulative Advantage*. In a broader economic context I shall refer to it here as the *Bestseller Effect*. Once a product gets a head start in sales, it signals to consumers that other people want that product and therefore it must be good, thereby causing them to desire it as well, which leads even more people to purchase the product, sending more signals to other consumers that they too must have it, and so it climbs up the bestseller list. Everyone in business knows about the effect, which is why authors and publishers, for example, try so fervently to land their book on the *New York Times* bestseller list. Once you are on the list, bookstores move your title to the "bestseller" bookcase (sometimes even labeled "New York Times Bestseller List") and to the front of the store where copies of the book are stacked like cordwood. This sends a signal to potential book buyers entering the store that this must be a good read, triggering an increase in sales that gets reported to the *New York Times Book Review* editors, who bump the title up the list, sending another signal to bookstore buyers to order even more copies, which secures the title more time on the bestseller list, which increases sales even further, and round and round the feedback loop goes as the richest authors get even richer.[3]

To quantify the Bestseller Effect, the Columbia University sociologist Duncan Watts and his collaborators Matthew Salganik and Peter Dodds tested it in a Web-based experiment in which fourteen thousand participants registered at a Web site where they had the opportunity to listen to, rate, and download songs by unknown bands.[4] One group of registrants were only given the names of the songs and bands, while a second group of registrants were also shown how many times the song had been downloaded. The researchers called this the "social influence" condition, because they wanted to know if seeing how many people had downloaded a song would influence subjects' decision on whether or not to download it. Predictably, the Web participants in the social influence condition were influenced by the download rate figures: songs with a higher download number were more likely to be downloaded by new participants, whereas subjects in the independent group who saw no download rates revealed dramatically different song preferences.[5] This is not to deny that the quality of a song or a book or any other product matters. Of course it does, and this too is measurable. But it turns out that subjective consumer preferences grounded in relative rankings by other consumers can and often do wash out the effects of more objective ratings of product quality.

Markets that traffic in rankings, ratings, and bestseller lists seem to operate on their own volition, almost like a collective organism. In fact, this is only one of many effects we shall see in this book that demonstrate just how much the mind influences the market, and in a broader sense how markets seem to have a mind of their own. Consider another economic parable with an evolutionary lesson related to the Bestseller Effect.

✦ ✦ ✦

Imagine that you are a banker with a limited amount of money to lend. If you advance loans to people who are poor credit risks, you are taking a great gamble that they will default on their loans and you will go out of business. This sets up a paradox: the people who most need the money are the worst credit risks and thus cannot get a loan, whereas the people who least need the money are the best credit risks, and thus once again the rich get richer. The evolutionary psychologists John Tooby and Leda Cosmides call this the *Banker's Paradox*, and they apply it to a deeper evolutionary problem: to whom should we extend our friendship? The Banker's Paradox, they suggest, "is analogous to a serious adaptive problem faced by our hominid ancestors: exactly when an ancestral hunter-gatherer is in most dire need of assistance, she becomes a bad 'credit risk' and, for this reason, is less attractive as a potential recipient of assistance."[6]

If we think of life as an economy, and if we count resources as anything we have that could help others—including and especially friendship—by the logic of the Banker's Paradox we have to make difficult choices in assessing the credit risk of people we encounter. In evolutionary theory the larger problem to be solved here is altruism: why should I sacrifice my genes for someone else's genes? Or, more technically, an altruistic act is one that lowers my reproductive success while simultaneously raising the reproductive success of someone else.

Standard theory suggests two evolutionary pathways to altruism: *kin selection* (blood is thicker than water) and *reciprocal altruism* (I'll scratch your back if you'll scratch mine). By helping my kin relations, and by extending a helping hand to those who will reciprocate my altruism, I am helping myself. Thus, there will be a selection for those who are inclined to be altruistic . . . to a point. With limited resources we can't help everyone and so we must assess credit risks, and some people are better risks than others. Here again is the Banker's Paradox: those most in need of assistance are the least likely to be given help, and so yet again the rich get richer. But not always, because fair-weather friends may be faking their signs of altruistic tendencies and later fail to come to our aid when the weather turns decidedly stormy. By contrast, true friends are those who are deeply committed to our welfare regardless of the potential for reciprocity. "It is this kind of friend that the fair-weather friend is the counterfeit of," Tooby and Cosmides continue. "If you are a hunter-gatherer with few or no individuals who are deeply engaged in your welfare, then you are extremely vulnerable to the volatility of events—a hostage to fortune."[7] The worse the environment, the more important it is that we have true friends, and the environment of our evolutionary past was no picnic.

Evolution, it is suggested, would have selected for adaptations to work around the Banker's Paradox dilemmas, including selecting us to (1) seek recognition from our fellow group members for our trustworthiness and reliability, (2) cultivate those attributes most desired by others in our group, (3) participate in social activities that recognize and reinforce such prosocial attributes, (4) avoid social activities that lead to untrustworthy actions and therefore a negative reputation, (5) notice similar attributes of trustworthiness in others, and (6) develop the ability to discriminate between true and fair-weather friends. Thus, Tooby and Cosmides conclude, the Banker's Paradox leads us to an evolved psychology where "if you are unusually or uniquely valuable to someone else—for whatever reason—then that person has an uncommonly strong interest in your survival during times of difficulty. The interest they have in your survival makes them, therefore, highly valuable to you. The fact that they have a stake in you means . . . that you have a stake in them. Moreover, to the extent they recognize this, the initial stake they have in you may be augmented."[8] Through such augmentation can the poor become rich by means of the evolved foundation of friendship.

✦ ✦ ✦

In 1859, Charles Darwin's *On the Origin of Species* was published. The book was so controversial that in 1861 the British Association for the Advancement of Science devoted a special session of its annual conference to it. Talks were given, pro and con, with one critic carping that Darwin's book was too theoretical and that he should have just "put his facts before us and let them rest." In attendance was Darwin's friend and colleague, the political economist and social activist Henry Fawcett, who wrote Darwin to report on the theory's reception (Darwin did not attend such meetings, usually due to ill health and family obligations). Darwin wrote Fawcett back, explaining the proper relationship between facts and theory:

> About thirty years ago there was much talk that geologists ought only to observe and not theorize, and I well remember someone saying that at this rate a man might as well go into a gravel-pit and count the pebbles and describe the colours. How odd it is that anyone should not see that all observation must be for or against some view if it is to be of any service![9]

This quote was the centerpiece of the first of my monthly columns for *Scientific American*, in which I elevated it to a principle I call *Darwin's Dictum*,[10] as identified in the final clause: *all observation must be for or against some view if it is to be of any service.*

Darwin's Dictum encodes the philosophy of science of this book: if observations are to be of any use, they must be tested against some view—a thesis, model, hypothesis, theory, or paradigm. Since the facts never just speak for themselves, they must be interpreted through the colored lenses of ideas; percepts need concepts. Science is an exquisite blend of data and theory—percepts and concepts—that together form the bedrock for the foundation of science, the greatest tool ever devised for understanding how the world works. We can no more separate our theories and concepts from our data and percepts than we can find a truly objective Archimedean point—a god's-eye view—of ourselves and our world.

One view that I am writing against in this book, ironically, is the belief that Darwin and the theory of evolution have no place in the social sciences, especially in the study of human social and economic behavior. Whereas scientists are up in arms about attempts to teach creationism and Intelligent Design in public school biology classrooms (see my book *Why Darwin Matters*), and are distraught over the dismal state of science education and the lack of acceptance of Darwin's theory (less than half of Americans believe that humans evolved),[11] most scientists—especially social scientists—have resisted with the emotional intensity of a creationist any attempts to apply evolutionary thinking to psychology, sociology, and economics. The original reason for this resistance—understandable at the time—was the equation of evolutionary theory with Social Darwinism and especially the extreme hereditarian views that led to enforced sterilization of the mentally retarded in America, and to the Nazi eugenics program that gave rise to the Holocaust. As a consequence, post–World War II social scientists steered a wide course around any attempts to apply evolutionary theory to the study of human behavior, and instead focused almost exclusively on sociocultural explanations.

A second view that I am writing against is the theory of *Homo economicus*, which holds that "Economic Man" has unbounded rationality, self-interest, and free will, and that we are selfish, self-maximizing, and efficient in our decisions and choices. When evolutionary thinking and modern psychological theories and techniques are applied to the study of human behavior in the marketplace, we find that the theory of *Homo economicus*—which has been the bedrock of traditional economics—is often wrong or woefully lacking in explanatory power. It turns out that we are remarkably irrational creatures, driven as much (if not more) by deep and unconscious emotions that evolved over the eons as we are by logic and conscious reason developed in the modern world.

A third view that I am writing against is the belief, first propounded in 1849 by the British historian Thomas Carlyle, that economics is "the dis-

mal science." For the next century and a half most people thought of it
that way, seeing only a field bogged down in mathematical models, financial analyses, and theoretical representations of people as rationally
calculating and maximally selfish machines. In reality, when we examine
all three of these views together, we find that economics is anything but
dismal. First, it is undergoing the most dynamic revolution since Adam
Smith founded the science in 1776 with his book *The Wealth of Nations*.
Rich transdisciplinary hybrids are emerging to breathe new life into an
old science, such as evolutionary economics, complexity economics, behavioral economics, neuroeconomics, and what I call *virtue economics*.
Second, and more important, people, companies, and nations care deeply
and passionately about their finances, and they always have. On this level,
economics has never been dismal. Put a couple of liberals and a couple of
conservatives in a room together and ask them to dispassionately discuss
the economics of universal health care, the privatization of social services,
the cost-benefits of foreign aid, or the relative merits of a flat tax versus a
progressive tax, and see just how quickly the tone of the conversation will
escalate into a state that is anything but dismal.

✦   ✦   ✦

I have spent thirty years in science dealing with such controversial topics as evolution, creationism, global warming, Holocaust denial, racial
differences in I.Q., racial differences in sports, gender differences in
cognitive abilities, conspiracy theories ranging from Pearl Harbor and
9/11 to the JFK, RFK, and MLK assassinations, alternative and complementary medicine, reincarnation and the afterlife, and even God
and religion. Yet it has been my experience that as ruffled feathers go,
economics is second to none in emotive volatility. If ever we need im-
partiality in our assessment of the facts—especially when the facts do
not just speak for themselves—it is in economics. We must study the
laws of human behavior in economics as the physicist, chemist, or biologist studies the laws of nature; and when we do so, because we are
dealing with a subject in which most people are emotionally invested,
we must make a ceaseless effort not to ridicule, bewail, or scorn human actions, but to understand them. Allow me to explain how I came
to this subject.

In the mid-1970s, I was an undergraduate at Pepperdine University, a Church of Christ institution with a strong conservative bent at a time when liberals ruled academe. I matriculated there because I was an evangelical Christian who wanted to be a college professor, so theology seemed like the most appropriate field and Pepperdine had a strong theology department (it didn't hurt that the campus is located in the majestic Malibu hills overlooking the Pacific Ocean). I soon discovered, however, that in order to earn a Ph.D. in theology one had to master four dead languages—Hebrew, Greek, Latin, and Aramaic—and since I found even Spanish taxing, this career choice was problematic. When my advisers also warned me about the dubious university job market for theologians, and my parents began to wonder aloud what I was planning to do for a living, I switched to psychology, where I discovered the language of science, in which I flourished. Theology is based on logical inquiry, philosophical disputation, and literary deconstruction. Science is founded on empirical data, statistical analysis, and theory building. For my style of thinking, the latter was a better fit.

My introduction to economics came in my senior year when many of the students in the psychology department were reading a cinder block of a book entitled *Atlas Shrugged*, by the novelist-philosopher Ayn Rand. I had never heard of the book or the author, and the novel's size was so intimidating that I refused to join the ranks of the enthused for months, until social pressure pushed me into taking the plunge. I trudged through the first hundred pages (patience was strongly advised) until the gripping mystery of the man who stopped the motor of the world swept me through the next thousand pages.

I found *Atlas Shrugged* to be a remarkable book, as many have. In fact, in 1991 the Library of Congress and the Book-of-the-Month Club surveyed readers about books that "made a difference" in their lives. *Atlas Shrugged* was rated second only to the Bible.[12] Rand's philosophy of Objectivism was based on four fundamental principles: (1) Metaphysics: Objective Reality; (2) Epistemology: Reason; (3) Ethics: Self-interest; (4) Politics: Capitalism.[13] Although I now disagree with her ethics of self-interest (science shows that in addition to being selfish, competitive, and greedy, we also harbor a great capacity for altruism, cooperation, and charity), reading Rand led me to the extensive body of literature on business, markets, and economics.

I cannot say for certain whether it was the merits of free market economics and fiscal conservatism (which are considerable) that convinced me of its veracity, or if it was my disposition that reverberated so well with its worldview. As it is for most belief systems we hold, it was probably a combination of both. I was raised by parents who could best be described as fiscally conservative and socially liberal. Products of the Depression and motivated by the fear of falling back into abject poverty, they skipped college and worked full-time well into their later years. Throughout my childhood I was inculcated with the fundamental principles of economic conservatism: hard work, personal responsibility, self-determination, financial autonomy, small government, and free markets. Even though they were not in the least religious (as so many conservatives are today), my parents were exceedingly generous to those who were less fortunate—greed is good, but charity is better.

After Pepperdine, I began a graduate program in experimental psychology at California State University, Fullerton, by which time I had abandoned my religious faith and embraced in its stead the secular values of the Enlightenment and the rigorous methods and provisional truths of science.[14] But after two years of enticing rats to press bars in proportion to the frequency and intensity of the reinforcements we gave them, my enthusiasm for practicing this type of science waned, while my wanderlust for the real world waxed.[15] I went to the campus career development office and inquired what I might do for a living with a master's degree. "What are you educated to do?" they inquired. "Train rats," I replied sardonically. "What *else* can you do?" they persisted. "Well," I searched, "I can research and write." The employment book included a job description for research and writing at a trade magazine of the bicycle industry, about which I knew nothing. My first assignment was to attend a press conference hosted by Cycles Peugeot and Michelin Tires in honor of John Marino, a professional bicycle racer who broke the transcontinental record from Los Angeles to New York. I fell in love with the sport, entering my first race that weekend, and for the next two years I learned the business of publishing, the economics of sales and marketing, and the sport of cycling. I wrote articles, sold advertisements, and rode my bike as far as I could. At the end of 1981, I left the magazine to race full-time, supported by corporate sponsors and an adjunct professor's salary from teaching psychology at Glendale College.

One day in 1981, during a long training ride, Marino told me about Andrew Galambos, a retired physicist teaching private courses through his own Free Enterprise Institute, under an umbrella field he called "Volitional Science." The introductory course was V-50. This was Econ 101 on free market steroids, an invigoratingly muscular black-and-white world where Adam Smith is good, Karl Marx bad; individualism is good, collectivism bad; free economies are good, mixed economies bad. The course was popular in Orange County, California (labeled by our neighbors in L.A. County as the "Orange Curtain"), and the time was right with Ronald Reagan as president and conservatives on the ascendant. Where Rand advocated for limited government, Galambos proffered a theory in which everything in society would be privatized until government simply falls into disuse and disappears. Galambos identified three types of property: *primordial* (one's life), *primary* (one's thoughts and ideas), and *secondary* (derivatives of primordial and primary property, such as the utilization of land and material goods). Thus, Galambos defined capitalism as "that societal structure whose mechanism is capable of protecting all forms of private property completely." To realize a truly free society, then, we have merely "to discover the proper means of creating a capitalist society." In this free society, we are all capitalists.[16]

Galambos's story is not unusual in the history of the oft-fringy libertarian movement. He had a massive ego that propelled him to a successful career as a private lecturer, but led him to such ego-inflating pronouncements as his classification of all sciences into physical, biological, and his own "volitional sciences." His towering intellect took him to great heights of interdisciplinary creativity, but often left him and his students tangled up in contradictions, as when we all had to sign a contract promising that we would not disclose his ideas to anyone, while we were also inveigled to solicit others to enroll. ("You've got to take this great course." "What's it about?" "I can't tell you.") And he had a remarkable ability to lecture for hours without notes in an entertainingly colloquial style, but when two hours stretched into three, and three hours dragged into four, his audiences were never left wanting for more. Most problematic, however, was any hope of translating theory into practice, which is where the rubber meets the road for any economic principle. Property definitions are all well and good, but what happens when we cannot agree on property rights infringements? The answer was inevitably something like this: "In a truly free society all such disputes will be peacefully resolved through private arbitration." This sounds good in theory and makes for a nice just-so story, but I would like more data from real world social experiments.

Galambos had a protégé named Jay Stuart Snelson, whom I met shortly after taking V-50. Snelson taught courses at the Free Enterprise Institute, but after a falling-out with Galambos (a common occurrence in Galambos's social sphere that also plagued Ayn Rand and other libertarian leaders), Snelson founded his own Institute for Human Progress. To distance himself from Galambos and bring his ideas more in line with mainstream economic theory, Snelson built on the shoulders of what is known as the Austrian School of Economics, most notably the work of the Austrian economist Ludwig von Mises. Snelson demonstrated through a series of economic principles and historical examples that free market capitalism is unquestionably the most effective means of optimizing peace, prosperity, and freedom, and that the privatization of education, transportation, communications, health services, environmental protection, crime prevention, and countless other areas would produce the greatest good for the greatest number.

During this time Marino and I (and our cycling partner Lon Haldeman) turned our cycling passion into a business called the Race Across America—a three-thousand-mile nonstop transcontinental bicycle race—with corporate sponsors and a contract from ABC Sports. Several appearances on *Wide World of Sports* gave me the recognition and confidence to open Shermer Cycles, a bicycle shop in Arcadia, California. Meanwhile, I expanded my teaching duties by creating new courses in evolutionary theory and the history of ideas at Glendale College.[17] I also developed a monthly seminar reading group called the "Lunar Society"—after the famous eighteenth-century Lunar Society of Birmingham—centered on discussing such books as *Human Action*, which inspired me toward the lofty goal set by its author, Ludwig von Mises: "One must study the laws of human action and social cooperation as the physicist studies the laws of nature." I call this *Mises's Maxim*, and it is one of two principles that guide my thinking in this book.[18]

In 1987, I decided that if I wanted to make an impact on the world through ideas I was going to have to give up my competitive cycling career and complete my graduate studies. I switched fields from psychology to history, and in 1991 I graduated from Claremont Graduate University with a Ph.D. in the history of science. I began teaching at Occidental College, a prestigious four-year liberal arts college in Los Angeles, and since I was interested in broader issues in science, particularly the growing threat of pseudoscience and irrationality in our culture, in 1992, I cofounded (along with my wife, Kim, and the artist Pat Linse) the Skeptics Society, *Skeptic* magazine, and our public science lecture series at the California Institute of Technology.

The motto of the Skeptics Society is the second guiding principle of this book, and it comes from the Dutch philosopher Baruch Spinoza's 1667 treatise on politics penned just before his death, *Tractatus Politicus*, in which he explained his methodology for studying such emotionally charged subjects as politics and economics:

> That I might investigate the subject matter of this science with the same freedom of spirit as we generally use in mathematics, *I have labored carefully not to mock, lament, or denounce human actions, but to understand them*; and to this end I have looked upon passions, such as love, hatred, anger, envy, ambition, pity, and the other perturbations of the mind, not in the light of vices of human nature, but as properties just as pertinent to it as are heat, cold, storm, thunder, and the like to the nature of the atmosphere.[19]

A pithier translation of the key phrase reads: "I have made a ceaseless effort not to ridicule, not to bewail, not to scorn human actions, but to understand them." I elevated it to *Spinoza's Proverb*, a standard toward which to reach when dealing with such emotionally laden topics as science, religion, and morality, which encompass my belief trilogy: *Why People Believe Weird Things*, *How We Believe*, and *The Science of Good and Evil*.[20] It is no less so with evolutionary economics. To that end, economics is for everyone.

1

THE GREAT LEAP FORWARD

Living along the Orinoco River that borders Brazil and Venezuela are the Yanomamö people, *hunter-gatherers* whose average annual income has been estimated at the equivalent of about $100 per person per year. If you walk into a Yanomamö village and count up the stone tools, baskets, arrow points, arrow shafts, bows, cotton yarn, cotton and vine hammocks, clay pots, assorted other tools, various medicinal remedies, pets, food products, articles of clothing, and the like, you would end up with a figure of around three hundred. Before ten thousand years ago, this was the approximate material wealth of every village on the planet. If our species is about a hundred thousand years old, then 90 percent of our history has been spent in this state of relative economic simplicity.[1]

Living along the Hudson River that borders New York and New Jersey are the Manhattan people, *consumer-traders* whose average annual income has been estimated at $40,000 per person per year. If you walk into the Manhattan village and count up all the different products available at retail stores and restaurants, factory outlets, and superstores, you would end up with a figure of around ten billion. This is a mind-boggling comparison, first made by the economist Eric Beinhocker in his comprehensive study *The Origin of Wealth*. Something happened over the last ten thousand years to increase the average annual income of hunter-gatherers by four hundred times.

As remarkable as this jump in income is, it pales in comparison to the differences between hunter-gatherers and consumer-traders in terms of product count, which in modern economics is measured in *Stock Keeping Units*, or *SKUs*, a retail measure of the number of types of products available in a store. By one estimate, for example, seven hundred new products are introduced into the market every day, a quarter of a million a year. In 2005, there were 26,893 new food and household products alone, including 187 new breakfast cereals, 303 new women's fragrances, and 115 new deodorants. Between the Yanomamö's three hundred SKUs and the Manhattans' ten billion SKUs is a difference of 33 million times.[2]

This difference of 400 times in income and 33 million times in products almost beggars description. We need analogies to get our minds around this staggering disparity. One contrast of the income difference: at its widest point, Manhattan island is 3.7 kilometers across, a distance you could easily walk in less than an hour while window shopping and skyscraper gazing. Multiply that figure by 400 and you get 1,480 kilometers, or slightly more than the distance from New York to Atlanta, which would take you 261 hours (10.9 days) to walk at a comfortable pace without stopping. Even more dramatic is a comparison of the SKU difference. The length of Manhattan is 21.5 kilometers. Multiply that by 33 million and you get a figure of 709,500,000 kilometers, or the approximate distance between Earth and Jupiter when both planets are in their orbits on the same side of the sun. You can walk the length of Manhattan in a day, but even traveling at the breakneck speed of a little over 51,000 kilometers per hour, it took the Voyager I spacecraft a year and a half to get to Jupiter.[3]

If ever there was a great leap forward, this was it, comparable to the evolution of bipedalism, the big brain, and consciousness, equivalent to the invention of fire, the printing press, and the Internet, and on par with the Agricultural Revolution, the Industrial Revolution, and the Digital Revolution. And this great leap did not happen gradually. It has been estimated that the $100 per person annual income had risen only to about $150 per person by 1000 BCE—the end of the Bronze Age and the time of King David—and did not exceed $200 per person per annum until after 1750 and the onset of the Industrial Revolution. In other words, it took 97,000 years to go from $100 to $150 per person per year, then another 2,750 years to climb to $200 per person per year, and, finally, 250 years to ascend to today's level of $6,600 per person per year for the entire world—and as we just saw, an order of magnitude higher still for the wealthiest people in the richest nations. If we compressed that 100,000-year period into just one year, then the last 250-year period of relative prosperity would represent less than one day out of the year. Or, if we condensed the hundred millennia into one 24-hour day, our epoch of industrial production and market economies accounts for a mere 3.6 minutes. In other words, the age in which we live and take for granted as normal and the way things have always been, in fact constitutes a mere one-quarter of one percent of the history of humanity.

How and why did humans make this great economic leap? We can answer this question by employing the methods and findings of science from a number of related and revolutionary new fields, including complexity theory, evolutionary psychology, evolutionary economics, behavioral economics, neuroeconomics, and virtue economics. We need all of these new fields—along with those from the traditional sciences—brought to bear on this question because it remains one of the greatest unsolved mysteries of our time.

For simplicity's sake, I shall lump all of these sciences under the rubric of *Evolutionary Economics—the study of the economy as an evolving complex adaptive system grounded in a human nature that evolved functional adaptations to survival as a social primate species in the Paleolithic epoch in which we evolved.* This is a swanky way of saying that the economy is a very complex system that changed and adapted to circumstances as it evolved out of a much simpler system, that we spent the first ninety thousand years of our life as hunter-gatherers living in small bands, and that this environment created a psychology not always well equipped to understand or live in the modern world. In essence, in trying to explain the Great Leap Forward, I am addressing three problems about the mind of the market:

1. *How the market has a mind of its own*—that is, how economies evolved from hunter-gathering to consumer-trading.
2. *How minds operate in markets*—that is, how the human brain evolved to operate in a hunter-gatherer economy but must function in a consumer-trader economy.
3. *How minds and markets are moral*—that is, how moral emotions evolved to enable us to cooperate and how this capacity facilitates fair and free trade.

This is a really hard problem to solve.

◆ ◆ ◆

Ever since I took a course in astronomy at the start of my college education, I have noticed a disturbing tendency on the part of both the scientific community and the culture at large to rank the sciences from "hard" (the physical sciences, such as astronomy, physics, and chemistry) to "medium" (the biological sciences, such as anatomy, physiology, and zoology) to "soft" (the social sciences, such as psychology, sociology, and anthropology). History was not even considered a science, and economics was off in its own land in the balkanized world of academia. And as rankings are wont to be, this hierarchy includes an assessment of worth, with the hard sciences being the most worthy and the soft sciences the least, accompanied by corresponding levels of recognition and support. Yet, having had some training in the physical and biological sciences, and extensive education and experience in the social sciences, I have always felt that the rank order is precisely reversed.

The physical sciences are hard in the sense that calculating differential equations is difficult, for example; but the subject matter itself is relatively easy to define and explain when compared to the considerably more complex and interconnected world of life and ecosystems. Yet even the difficulty of constructing a comprehensive theory of biology—which still remains a knotty problem in the life sciences—is pallid in comparison to that of the workings of human brains and societies. In my opinion, the social sciences are the hard sciences, because our subject matter is orders of magnitude more complex and multifaceted.

In the neurosciences, the study of consciousness has come to be known as "the hard problem"[4] because it has been so difficult to explain how the activity of billions of individual neurons generates the collective phenomenon of conscious thought, or what one scientist calls the "society of mind."[5] An even harder problem—what I call *the really hard problem*—is for science to explain how the activity of billions of individual humans generates the collective phenomenon of *culture,* or the "society of culture," and what its proper economic and political structure should be to achieve social harmony.

As humans made the transition from hunter-gatherers to consumer-traders, groups have tried hundreds of different social experiments in an attempt to solve the really hard problem. Bands, tribes, chiefdoms, states, and empires have been formed. Theocracies, plutocracies, monarchies, and democracies have been tried. Tribalism, statism, socialism, and now globalism have been practiced. From no trade to fair trade to free trade, endless permutations of economic arrangements have been employed, with greater and lesser success. And for millennia, philosophers and scholars from all walks of life and from around the world have attempted to solve the really hard problem with little consensus. Can modern science do better?

WRONG

✦ ✦ ✦

Evolution is a complex system that emerges out of the simple actions of organisms just trying to survive and provide for their offspring. Economies are complex systems that emerge out of the simple actions of people just trying to make a living and provide for their children. Thus, when we address (1) how economies evolved from hunter-gathering to consumer-trading, (2) how the human brain evolved to operate in a hunter-gatherer economy but must function in a consumer-trader economy, and (3) how moral emotions evolved to enable us to cooperate and how this capacity enables fair and free trade, we are really studying (1) the behavior of markets and economies, (2) the psychology of people operating in markets and economies, and (3) the moral aspects of markets and economies.

Evolution and economics are not just analogous to one another; they are actually two different examples of a larger phenomenon called *complex adaptive systems,* in which individual elements, parts, organisms, or people interact, process information, and adapt their behavior to changing *how?*

conditions. These are systems that learn and grow as they evolve from simple to complex, and they are *autocatalytic*, which means that they contain self-driving feedback loops (as when a PA system develops a feedback loop between speakers and microphone, resulting in an accelerating rise in pitch and volume). Here are some examples of complex adaptive systems and what emerges from them, as the built-in autocatalysis leads to their being self-organized:

TELEOLOGICAL

Life is a self-organized emergent property of prebiotic chemicals that came together in a manner that allowed them to be self-sustaining and capable of duplication and reproduction.

Complex Life is a self-organized emergent property of simple life, as when simple prokaryote cells coalesced into the more complex eukaryote cells of which we are made, which contain within them organelles that were once prokaryote cells (such as mitochondria, which have their own DNA).

Multicellular Life is a self-organized emergent property of single-celled life forms, which merged together as a cooperative strategy for more successful survival and reproduction.

Immunity is a self-organized emergent property of billions of cells of our immune system working together to combat bacteria and viruses.

Consciousness is a self-organized emergent property of billions of neurons firing in complex patterns in the brain.

Language is a self-organized emergent property of thousands of words spoken in communication among language users.

Law is a self-organized emergent property of thousands of informal mores and restrictions that were codified over time into formal rules and regulations as societies grew in size and complexity.

Economy is a self-organized emergent property of millions of people pursuing their own self-interests with little awareness of the larger complex system in which they work.

Complex adaptive systems appear designed from the top down, but in fact as they evolve they construct objects from the bottom up through *functional adaptations*—what works survives and reproduces into the future designscape of life or culture. From the simplest forms of life up the chain of complexity, we move from simple cells to complex cells to multicellular organisms, to colonies, to social units, to societies, to consciousness, to language, to the law, and to economies.

In living organisms, the complex adaptive system of evolution is driven by natural selection, or *variation plus cumulative selection*. The proverbial monkey randomly pecking away at the computer keyboard will not produce Hamlet in a billion years, or even "To be or not to be" any time soon. But add a nonrandom cumulative selection component to the equation that preserves the correct letters and erases the typos, and

wieTskewkOsdfeB92uE2OseRdl7jeNkseOdseTe3r22TsweOsxB wxseE . . .

becomes

TOBEORNOTTOBE . . .

In fact, in order to demonstrate the power of cumulative selection, my friend and colleague Richard Hardison designed a computer program to do precisely this, in which randomly typed letters were "selected" for or against this standard, taking less than ninety seconds and only 335.2 trials to produce the famed Shakespearean soliloquy that by chance would have taken 26^{13} number of trials to produce. Simultaneous with and independent of Hardison, Richard Dawkins designed a similar cumulative selection program for a different Shakespearean soliloquy, METHINKSITISLIKEAWEASEL. As Dawkins noted when we later discovered the coincidence,

Once one has grasped (from Darwin) the paramount importance of ratcheted *cumulative* selection when faced with the Argument from Statistical Improbability, one's thoughts naturally turn to the famous monkeys who have so often been used to dramatise that Argument. It becomes the obvious simulation to do, to get the point across to doubters. It can easily be done with a little BASIC program, and that is what both Hardison and I did, at what must have been almost exactly the same time, 1984 or 1985. As for the superficial details, those pesky monkeys have always typed

Shakespeare. *Hamlet* is his most famous play. "To be or not to be" is the most famous passage from that play. I would probably have chosen it myself, except that I thought the dialogue between Hamlet and Polonius on chance resemblances in clouds would make a neat intro: hence "Methinks it is like a Weasel."[6]

In nature, random genetic mutations and the mixing of parental genes in offspring produce variation, and the *selection* of this genetic variation through the survival of their hosts is what drives evolution. Out of this process of self-organized directional selection emerge complexity and diversity. This is how evolution is a self-organized emergent phenomenon.

The evolution of our material economy proceeds in an analogous manner through the production and selection of numerous permutations of countless products. Those ten billion products in the Manhattan village represent only those variations that made it to market, so there is already a selection process by the manufacturers themselves as they attempt to correctly predict what the market will prefer. Once these choices are brought to market, there is a cumulative selection for those deemed most useful or desired, with the selection conducted by consumers in the marketplace who vote with their dollars on which will survive: VHS over Betamax, DVDs over VHS, CDs over records, flip phones over brick phones, computers over typewriters, Google over Altavista, SUVs over station wagons, paper books over e-books (still), and Internet news over network news (soon).[7] Those that are purchased "survive" and "reproduce" into the future through repetitive use and remanufacturing.

The environment is the designspace of evolution. The market is the designspace of economics. Just as nature selects the variation best suited to survive in a particular environment, so too do people select the goods and services that are best suited to meet their unique needs and desires in a particular market. Note that in neither evolution nor economics is there a top-down designer to oversee the entire system. In life, no one is "selecting" organisms for survival or extinction, in the benign sense of animal breeders or in the malignant sense of Nazi doctors. Evolution is unconscious and nonprescient—it cannot look forward to anticipate what changes are going to be needed for survival. The analogy with the economy is not perfect, since some top-down institutional rules and laws are needed to provide a structure within which free and fair trade can occur. But too much top-down interference with the marketplace makes trade neither free nor fair, and when such attempts have been made in the past they have failed because markets are far too complex, interactive, and autocatalytic. In his 1922 book, *Socialism*, Ludwig von Mises spelled out the reasons why, most notably the problem of "economic calculation" in a planned socialist economy. In capitalism, prices are in constant and rapid flux and are determined from below by individuals freely exchanging in the marketplace; in socialism, prices are slow to change and are determined from above by government fiat. Money is a means of exchange and prices are the information people use to guide their choices. Mises demonstrated that socialist economies depend on capitalist economies to determine what prices should be assigned to goods and services. And they do so cumbersomely and inefficiently. Ultimately, free markets are the only way to find out what buyers are willing to pay and what sellers are willing to accept.[8]

For example, studies show that Internet airfares change thousands of times an hour as people search for the best price they can find to reach their destinations. Airlines have sophisticated software programs that adjust the prices according to supply and demand for particular routes, the number of seats available at any given moment, and other variables that go into what has become known as "dynamic pricing."[9] Imagine a centralized bureaucratic airline price committee meeting each morning to work out how much it is going to cost someone to fly, say, from Greensboro, North Carolina, to Wichita, Kansas, on each of a dozen different airline carriers, factoring in not only the real-time supply and demand parameters and number of available seats, but also the time of day, type of aircraft, class of travel, cost of aviation fuel, number of frequent flyer mileage seats already taken, discount coupons, and dozens of other variables, and doing so for hundreds of thousands of people. It is not an impossible feat to attempt, and planned economies have tried to do it, but like the proverbial bipedal dog, it is exceedingly rare, clumsy, and painfully humorous to watch. Unfortunately for those forced to live in planned economies in the twentieth century, the economic disasters that were the inevitable result were neither rare nor humorous.

One more example will suffice to make my point. Imagine the futility of government bureaucrats trying to find the right price for each of the approximately 170,000 different books published each year and available online at Amazon.com, BarnesandNoble.com, Buy.com, eBay, Half.com, and so on, factoring in hardback versus paperback prices, special discounts for multiple purchases of bundled books, plus shipping specials for minimum sales, and factoring in, of course, the discriminatory pricing now used in the same way the airlines price their tickets, and then imagine multiplying that process by the hundreds of thousands of different markets, industries, and businesses, and it becomes crashingly clear why no top-down system could ever match the real-time sensitivity to prices provided by the bottom-up complex adaptive pricing system currently in place. Expand the problem by many orders of magnitude and we get a sense of the breathtaking inanity of trying to control an entire economy. Only millions of buyers and sellers in constant and real-time negotiation with millions of other buyers and sellers can determine the prices of those ten billion products, along with the hundreds of millions of services offered in a modern economy.

As with living organisms and ecosystems, the economy looks designed, so we naturally infer that a top-down designer (government) is needed in nearly every aspect of the economy. But just as living organisms are designed from the bottom up by natural selection, so too is the economy designed from the bottom up by the "invisible hand."

✦ ✦ ✦

Because the first ninety thousand years of our history were spent as hunter-gatherers living in small bands of a few dozen to a few hundred people, we evolved a psychology not always well equipped to reason our way around the modern world. What may seem like irrational behavior today may actually have been rational a hundred thousand years ago. Without an evolutionary perspective, the assumptions of *Homo economicus* make no sense. Take economic profit versus psychological fairness as an example.

Behavioral economists employ an experimental procedure called the Ultimatum Game. It goes something like this. You are given $100 to split between yourself and your game partner. Whatever division of the money you propose, if your partner accepts it, you are both richer by that amount. How much should you offer? Why not suggest a $90-$10 split? If your game partner is a rational, self-interested money-maximizer—as predicted by the standard economic model of *Homo economicus*—he isn't going to turn down a free ten bucks, is he? He is. Research shows that proposals that deviate much beyond a $70-$30 split are usually rejected.[10]

Why? Because they aren't fair. Says who? Says the moral emotion of "reciprocal altruism," which evolved over the Paleolithic eons to demand fairness on the part of our potential exchange partners. "I'll scratch your back if you'll scratch mine" only works if I know you will respond with something approaching parity. The moral sense of fairness is hardwired into our brains and is an emotion shared by all people and primates tested for it. Thousands of experimental trials with subjects from Western countries have consistently revealed a sense of injustice at lowball offers. Further, we now have a sizable body of data from peoples in non-Western cultures around the world, including those living close to the way our Paleolithic ancestors lived, and although their responses vary more than those of modern peoples living in market economies, they still show a strong aversion to unfairness.[11]

The deeper evolution of this can be seen in the behavior of our primate cousins. In studies with both chimpanzees and capuchin monkeys, the Emory University primatologist Frans de Waal found that when one individual is rewarded for work with a desired food and does not share it with a task partner, the partner refuses to cooperate in the future and expresses displeasure at the injustice.[12] Such results suggest that all primates, including and especially humans, evolved a sense of justice, a moral emotion that signals to the individual that an exchange was fair or unfair. Overwhelming evidence from numerous fields now shows that fairness evolved as a stable strategy for maintaining social harmony in our ancestors' small bands, where cooperation was reinforced and became the rule while freeloading was punished and became the exception. Apparently irrational economic choices today—such as turning down a free $10 with a sense of righteous injustice—were at one time rational when seen through the lens of evolution.

Just as it is a myth that evolution is driven solely by "selfish genes" and

FALSE

that organisms are exclusively greedy, self-centered, and competitive, it is a myth that the economy is driven exclusively by selfish intentions, and that people are exclusively greedy, self-centered, and competitive. The fact is, we are both selfish and selfless, cooperative and competitive, peaceful and bellicose, prosocial and antisocial. There exist in both life and economies mutual struggle and mutual aid. In the main, however, the balance in our nature is heavily on the side of good over evil. For every random act of violence that makes the evening news, there are ten thousand nonrandom acts of kindness that go unrecorded every day. Markets are moral, and modern economies are founded on our virtuous nature. If this were not the case, market capitalism would have imploded long ago.

NOPE

This is not a Pollyannaish view of economics, and I am not arguing that businesses operating in a free market vacuum are necessarily always moral. We need checks and balances and a society based on the rule of law within which markets can be both free and fair. James Madison was right when he wrote in Federalist Paper Number 51, "If men were angels, no government would be necessary. If angels were to govern men, neither external nor internal controls on government would be necessary."[13]

✦ ✦ ✦

Why do the sciences of complex systems and human nature predict and demonstrate that robust goodness will emerge from chaos and selfishness? As a social primate species, we evolved to display within-group amity and between-group enmity, which leads to a fierce tension between our selfish desire for individual gain or family unity and our social desire for group equality or social unity against outside threats. The first principle of what I call *virtue economics* is this genetically embedded reflex of reciprocity: when someone gives us something, we feel that we should give something back.

For both within-group and between-group interactions, we are greatly shaped by our perceptions of others, especially our perception of what others think of us; that is, we care about our reputation and status. This is why reputation metrics have arisen so quickly on the Internet as a self-organized emergent property of trust. Ratings of sellers' reputations on eBay, rankings of the quality of reviewers' assessments of books on

Amazon, and the quality and quantity of connections between users as a measure of reputation in social and professional networking sites such as MySpace, Facebook, and LinkedIn are just a few examples of the need for trust in exchanges of any kind.

We want to be thought of as fair and honest dealers. Yet we are also exceptionally tribal, and group identity is essential to our sense of self. The unfortunate byproduct of such in-group/out-group tribalism is xenophobia. We have a natural aversion to Others, and we show a remarkable ability to sort people into in-group/out-group categories on the most minute levels of criteria—think of such gangs as the Crips and the Bloods, or such ethnic disputes as those between the Hutus and the Tutsis, the Albanians and the Serbs, or the Shiites and the Sunnis. Although we have legislated and educated these ancient tribal rituals out of our culture, their psychological underpinnings are still buried deep in our Paleolithic brains, waiting to be stirred into action. Sometimes they are stirred disastrously in the service of politics (and war), sometimes they are stirred competitively in the service of economics (and trade).

At the same time, we evolved from bands and tribes to chiefdoms and states, and communities struggled to find the right balance between freedom and equality by experimenting with different social technologies, leading to a shift from the equal distribution of wealth among bands to the emergence of hierarchical wealth as a token of status and power among tribes. The hunter-gatherer principle of egalitarianism (or at least the pretense of it) disintegrates as bands and tribes coalesce into chiefdoms and states. When wealth becomes a symbol of power, the value of virtue emerges to offset the rival value placed upon personal gain.

So much
bullshit.

2

OUR FOLK ECONOMICS

In 1873, thirteen years after the legendary confrontation over the theory of evolution between the Anglican bishop Samuel Wilberforce ("Soapy Sam") and the evolutionary biologist Thomas Henry Huxley ("Darwin's bulldog"), Wilberforce died in an equestrian fall. Of Wilberforce's tragic end Huxley quipped to the physicist John Tyndall, "For once, reality and his brain came into contact, and the result was fatal."

When it comes to such basic forces as gravity and such fundamental phenomena as falling, our intuitive sense of how the physical world works—our folk physics—is reasonably sound. Thus, even children get the humor of cartoon physics where, for example, a character runs off a cliff but does not fall until he realizes that he has left terra firma. (This phenomenon is known as *Coyotes interruptus*, in honor of Wile E. Coyote, who frequently fell to his doom in this manner while chasing his roadrunner nemesis.) But much of physics is counterintuitive—ranging from quantum mechanics in the micro world to global general relativity in the macro world—as is the case in many other disciplines as well, and before the rise of modern science we had only our folk intuitions to guide us.

Folk astronomy, for example, told us that the world is flat, celestial bodies revolve around the earth, and the planets are wandering gods who determine our future. Folk biology intuited an élan vital flowing through all living things, which in their functional design were believed to have been created ex nihilo by an intelligent designer. Folk psychology compelled us to search for the homunculus in the brain—the ghost in the machine—a mind somehow disconnected from the brain.

The reason folk science so often gets it wrong is that we evolved in an environment radically different from the one in which we live. We evolved in what the evolutionary biologist Richard Dawkins calls "Middle World"—a land midway between large and small, fast and slow, young and old.[1] Out of literary preference, I call it *Middle Land*. In the Middle Land of space, our senses evolved for perceiving objects of middling size—between, say, ants and mountains. We are not equipped to perceive bacteria, molecules, and atoms on one end of the scale, and quasars, galaxies, and expanding universes on the other end. In the Middle Land of speed, our eyes are adroit at detecting objects moving at a walking or running pace, but the slow growth of mountains and the movement of continents on one end of the scale and the fast speed of light on the other end, for all intents and purposes, are imperceptible. In the Middle Land of time, we live a scant three score and ten years, far too short a time to witness evolution, continental drift, or long-term environmental changes. This is one reason why it is so difficult for so many people to accept the theory of evolution or the science of global climate change. Evolution and climate change are counterintuitive, in the sense of representing change on a scale that we are unaccustomed to dealing with in our Middle Land lives.

Causal inference in folk science—that is, determining cause-and-effect relationships in the real world—is equally untrustworthy. We correctly surmise designed objects such as stone tools to be the product of an intelligent designer, and thus naturally assume that all functional objects in nature, such as eyes, must have been similarly intelligently designed. Lacking a cogent theory of how neural activity gives rise to consciousness, we imagine mental spirits floating within our heads.

+ + +

Since its inception as a science, economics has been mired in controversy over how to apply its data and theory to devise prescriptive solutions for descriptive problems. Market solutions to social problems are generally received with skepticism. Businessmen are distrusted, corporations looked at askance. There is also a well-known resentment against those who have most benefited from markets. This distrust and antipathy have their roots in folk science and in the limitations of our Middle Land intuitions about markets and the economy. Folk economics leads us to disdain excessive wealth, label usury a sin, and mistrust the invisible hand of the market. What we do not understand we often fear, and what we fear we often loathe. (As one *New Yorker* cartoon featuring two people in conversation reads: "I hated Bill Gates before it became so fashionable.")

The reason for folk economics is that in the small bands of hunter-gatherers in which we evolved there were no capital markets, no economic growth, no accumulation of wealth, no excessive disparity between rich and poor, very little division of labor, very little concentration of labor (most hunter-gatherers were multitaskers), and certainly no invisible hand at work. To see where the Middle Land of folk economics evolved and why it tasks us today, the following chart presents a rough approximation of our evolution from hunter-gatherers to consumer-traders:

THE EVOLUTION OF HUMAN GROUPS

Time Before Present	Group	Number of Individuals
100,000–10,000 years	Bands	10s–100s
10,000–5,000 years	Tribes	100s–1000s
5,000–3,000 years	Chiefdoms	1000s–10,000s
3,000–1,000 years	States	10,000s–100,000s
1,000–present	Empires	100,000s–1,000,000s

In this evolutionary time line there is an economic transition from the equal distribution of economic wealth among bands to the emergence of hierarchical wealth as a token of status and power among tribes; egalitarianism falls apart as bands and tribes coalesce into chiefdoms and states. Thus, in the modern world, a tension arises between our selfish desire to gain greater wealth and our social desire for equality (or at least that no one should be inordinately unequal—either too rich or too poor). In monstrously large modern states we have both abject poverty and unimaginable wealth, which causes considerable consternation. In most nations this translates into political policy to raise the poor and lower the rich, because during our evolutionary tenure we lived in a zero-sum (win-lose) world, in which one person's gain meant another person's loss. This is why reciprocity and food sharing are so important to hunter-gatherer peoples, and why they evolved customs and mores to enforce the sharing of the products of communal efforts at hunting and gathering.

The Ache of eastern Paraguay, for example, are full-time nomadic hunter-gatherers whose men make extended foraging excursions. The meat from hunting kills is shared widely within the tribe, and the individual hunter who actually makes the kill does not own it. However, the anthropologists Kim Hill and Hillard Kaplan discovered that the most successful hunters enjoyed greater access to women and reproductive success in the form of more surviving offspring. Further, Hill and Kaplan noted that because participation in the hunting groups was fluid and uncertain, the best hunters were encouraged to go on expeditions via offers of such practical and social rewards as child care and higher status in the group. And the food sharing and reward system was most prevalent in game hunting, which carries a low probability of success, whereas the highly predictable and readily available products of gathering are shared only within the nuclear family.[2] In like manner, the Inuit developed a system whereby they award the prized upper half of a polar bear's skin to the hunter who made the first strike on the bear with his spear. Since the long mane hairs of the bear skin are used to line women's boots, the hunter thus receives additional recognition at home from the recipient of his prize.[3] Men like to impress women and seek status and recognition from their male peers, and this fact can be exploited to the benefit of the social group.

Today, however, we live in a nonzero world in which improved science and technology has increased productivity to the point where we can generate ever-increasing amounts of food from the same or even fewer resources. But our brains operate as if we are still living in that zero-sum Middle Land.[4]

♦ ♦ ♦

The great historian Arnold Toynbee proclaimed that *"The Wealth of Nations* and the steam engine destroyed the old world and built a new one." The Nobel laureate economist George J. Stigler called *The Wealth of Nations* "the most important substantive proposition in all of economics."[5]

If Adam Smith's theory of economics is so profound and proven, why do some people reject it, as others reject the theory of evolution? Natural selection and the invisible hand—evolution and economics—are not religious tenets that one swears allegiance to or believes in as a matter of faith; they are factual realities of the empirical world, and just as one would not say "I believe in gravity," one should not proclaim "I believe in markets." The resistance to accepting free market economics involves specific social and psychological factors.

Because humans evolved in small groups of a few dozen to a few hundred individuals in hunter-gatherer communities, in which everyone was either genetically related or knew one another intimately, most resources were shared, wealth accumulation was almost unheard of, and excessive greed and avarice were punished. Thus we naturally respond to a free market system in which conspicuous wealth is paraded as a sign of success with envy and anger, and the expectation is that someone or something more powerful than those greedy individuals should implement corrective action. Call it *evolutionary egalitarianism*. Further, throughout most of the history of civilization, economic inequalities were not the result of natural differences in drive and talent between members of a society equally free to pursue their right to prosperity; instead, a handful of chiefs, kings, nobles, and priests exploited an unfair and rigged social system to their personal benefit and at the cost of impoverishing the masses. Our natural response is to perceive such inequalities as ill-gotten gains and to demand controls from the top down to limit the amount of wealth accumulated by any one individual. Whenever anyone says "They should do something about it," the *they* that is invoked is inevitably the social institution with the most power—in our case, the government.

Humans also have an amazingly low tolerance for economic ambiguity. Free markets are chaotic and uncertain, uncontrollable and unpredictable. Most of us have a low tolerance for such environments, and we have learned to expect that social institutions such as the government will bring a level of certainty to society. Earthquakes and hurricanes are called *acts of God*, and over the past century we have come to rely on *acts of Government* to make the necessary adjustments and provide comforting security, especially when we have not done so for ourselves. People who cannot afford (or who choose not to purchase) insurance against such acts of God typically expect government agencies such as the Federal Emergency Management Agency (FEMA) to rescue them when the risk reaches reality.

The economist Daniel Klein recounts a poignant example of the influence of the God of Government on economic policy, even on the policy proposals of a Nobel laureate economist. In a session of the 1995 annual meeting of the American Economic Association, Klein asked the Nobel laureate MIT economist Robert Solow why he did not favor school vouchers, which apply free market principles and mechanisms to public education. Solow responded, "It isn't for any economic reason; all the economic reasons favor school vouchers. It is because what made me an American is the United States Army and the public school system." Klein suggests that people resist free market economics because of their "tendency to see and love government as a binding communitarian force." Klein calls this *The People's Romance*, derived from a shared use of government services and a willingness to let the government define the boundaries of our in-groups. "Government creates common, effectively permanent institutions, such as the streets and roads, utility grids, the postal service, and the school system. In doing so, it determines and enforces the setting for an encompassing shared experience—or at least the myth of such experience."[6] Anyone who is skeptical of a government solution to an economic or social problem, then, is easily labeled as an out-group member, a rebel without a collective cause.

For most people, economic beliefs arise out of tribal, political commitments—liberals are supposed to be against an unregulated free market, conservatives for it—folk economics of another form. Yet both liberals and conservatives endorse the practice of hefty economic regulations and big government; they differ only on how much regulation should be passed and upon whom it should be imposed. Liberals want corporations regulated and government in the boardroom, whereas conservatives want a big military and government in the bedroom. Liberals call attention to fiscal abuse and budgetary malfeasance in the military, and yet as my friend and colleague David B. Schlosser, a businessman and congressional candidate from Arizona, points out, "Liberals like to think that assigning a 'good' task to the government (health care, for example) automatically imbues the entire process with an inherent effectiveness. They ignore the fact that the government employees making decisions for the tasks they favor are the same people whom liberals do not trust to make good decisions for themselves. Why would we want to put medicine into the hands of the same bureaucracy that pays $800 for Army wrenches and $2000 for Air Force toilet seats?"[7] As the social commentator and political humorist P. J. O'Rourke quipped: "If you think health care is expensive now, wait until you see what it costs when it's free!"

But tribal conservatives are not off the hypocritical hook, most notably in their claimed desire for less government in economic policy—except for government subsidies for big business. Ralph Nader correctly calls this "corporate welfare," and abuses can be found in the oil industry, large swaths of agriculture and farming, and especially defense contracting. One might argue that such subsidies are good for America, but they inevitably result in distortions in the decisions that consumers would normally make in the absence of such interventions into the economy. Conservatives also abandon free market principles and turn tribal when it comes to foreign trade, believing in the zero-sum ideology that producing a product cheaper overseas is not a gain for American consumers in the form of cheaper products, but a loss for American jobs and manufacturing.[8] Under every administration of the past century—Democratic and Republican—the government has grown in size along with the taxes needed to pay for it. And nearly everyone still holds to the folk belief that in order for our economy to be healthy, it must be heavily regulated from the top down.

◆ ◆ ◆

Yet the single most common myth found in objections to both the theory of evolution and free market economics is based on the presumption that animals and humans are inherently selfish and that the economy is like Tennyson's memorable description of nature: "red in tooth and claw." After the *Origin of Species* was published, the British philosopher Herbert Spencer immortalized natural selection in the phrase "survival of the fittest," one of the most misleading descriptions in the history of science and one that has been embraced by social Darwinists ever since, who apply it inappropriately to racial theory, national politics, and economic doctrines. Even Darwin's bulldog, Thomas Henry Huxley, reinforced what he called this "gladiatorial" view of life in a series of essays, describing nature as a system "whereby the strongest, the swiftest, and the cunningest live to fight another day."[9]

This view of life need not have become the dominant one.[10] In 1902, the Russian anarchist and social commentator Pyotr Kropotkin published his rebuttal to Spencer and Huxley in his book *Mutual Aid*. Calling out Spencer by phrase, for example, Kropotkin notes: "If we . . . ask Nature: 'who are the fittest: those who are continually at war with each other, or those who support one another?' we at once see that those animals which acquire habits of mutual aid are undoubtedly the fittest. They have more chances to survive, and they attain, in their respective classes, the highest development of intelligence and bodily organization." In numerous trips to the wild hinterlands of Siberia, Kropotkin discovered that animal species there were highly social and cooperative in nature, an adaptation for survival that he deduced played a vital role in evolution. "In the animal world we have seen that the vast majority of species live in societies, and that they find in association the best arms for the struggle for life: understood, of course, in its wide Darwinian sense—not as a struggle for the sheer means of existence, but as a struggle against all natural conditions unfavourable to the species." The same is true with human communities, he continued, noting the evidence of mutual aid in "savages," "barbarians," medieval towns, and even modern societies. "The mutual protection which is obtained in this case, the possibility of attaining old age and of accumulating experience, the higher intellectual development, and the further growth of sociable habits, secure the maintenance of the species, its extension, and its further progressive evolution. The unsociable species, on the contrary, are doomed to decay."

Kropotkin may have been an anarchist, but he was no crackpot when it came to human nature. "There is an immense amount of warfare and extermination going on amidst various species," he admitted, noting that "the self-assertion of the individual" is the other "current" in our nature that must be recognized. However, he added, "there is, at the same time, as much, or perhaps even more, of mutual support, mutual aid, and mutual defense. . . . Sociability is as much a law of nature as mutual struggle."[11]

It is a matter of balancing these dual currents of selfishness and selflessness, competition and cooperation, greed and generosity, mutual struggle and mutual aid. That this view of life was eclipsed by that of Spencer and Huxley probably has to do with where they were developed: the more competitive economy of the United Kingdom versus the more egalitarian economy of Russia.[12] Since Adam Smith was Scottish, he has long been associated with the selfish/competitive view of life, as commentators routinely discount (or don't even know about) his earlier work on the moral sentiments in which he argued that people are also social, empathetic, and cooperative.

Life is intricate and complex and looks intelligently designed, so our folk science intuition leads us to infer that there must be an intelligent designer. Analogously, economies are intricate and complex and look designed, so our natural inclination is to infer that we need an intelligent designer. The God of Government is thus considered the intelligent designer of our economic systems.

Yet life and economies are not intelligently designed from the top down; they arise spontaneously out of simpler systems from the bottom up. The explanation for this design may be found in the sciences of emergence and complexity theory, in which complex systems arise from simple systems. Life and economies, like language, writing, the law, civilizations, and cultures, all arise spontaneously as self-organized emergent properties from within systems themselves and without the aid of a blueprint design by a clever engineer. Neither God nor Government is needed to explain such phenomena. In their stead, natural selection and the invisible hand explain precisely how individual organisms and people, pursuing their own self-interest in their struggle to survive and make a living, generate the emergent property of complex ecologies and economies. As we shall see, ecologies and economies are *Complex Adaptive Systems* (CAS): systems in which individual particles, parts, or agents interact, process information, learn, and adapt their behavior to changing conditions.

An ecology is a complex adaptive system that evolved to solve the problem of how so many unrelated organisms and species in large biological communities can coexist in relative harmony. An economy is a complex adaptive system that evolved to solve the problem of how so many unrelated strangers in large cities can coexist in relative harmony. Charles Darwin and Adam Smith, each in his unique way trying to solve a specific problem, together stumbled across an elegant solution to what turns out to be a larger, overarching phenomenon: the emergence of complexity out of simplicity. The solution hinges on our evolved natures as individuals coupled to the nature of evolutionary systems and how they behave as collections of individuals acting on their evolved natures, or the evolution of complex adaptive systems.

3

BOTTOM-UP CAPITALISM

In October 1825, Charles Darwin matriculated at Edinburgh University for medical studies, following in the footsteps of his ambitious physician father, Robert, and his famous polymath grandfather, Erasmus. As a matter of general course curriculum, Darwin studied the works of the great Enlightenment thinkers, including David Hume, Edward Gibbon, and Adam Smith. A decade later, upon his return home from the five-year voyage around the world on HMS *Beagle*, Darwin revisited these works, reconsidering their implications in light of the new theory he was developing.[1] Although Darwin does not refer to Smith directly, Darwin scholars largely agree that he modeled his theory of natural selection after Smith's theory of the invisible hand.[2]

All great works of science are written for or against some particular view, and the view Darwin was writing against was that presented in the 1802 book *Natural Theology: or, Evidences of the Existence and Attributes of the Deity, Collected from the Appearances of Nature* by William Paley. Darwin knew this book so well that he once noted in a private letter: "I do not think I hardly ever admired a book more. . . . I could almost formerly have said it by heart." It is from Paley that we get the famous "watchmaker" argument for God's existence, so popular among Intelligent Design creationists today.

If Darwin was arguing with Paley, with whom was Paley arguing? In part, the Scottish economist Adam Smith. In Paley's argument for divine providence in nature, he starts by describing a breeding pair of sparrows, who are unaware of the long-term and unintended consequences of their act of reproduction—the survival of the species:

> When a male and female sparrow come together, they do not meet to confer upon the expediency of perpetuating their species. . . . They follow their sensations; and all those consequences ensue, which the wisest counsels could have dictated, which the most solicitous care of futurity, which the most anxious concern for the sparrow world, could have produced. But how do these consequences arise? . . . Those actions of animals which we refer to as instinct, are not gone about with any view to their consequences . . . but are pursued for the sake of gratification alone; what does all this prove, but that the prospection, which must be somewhere, is not in the animal, but in the Creator?

Behind the scenes, God is pulling the strings. And to describe this process, Paley invokes the metaphor of the invisible hand, but inverts it from Smith's bottom-up process to a top-down divine force:

> For my part, I never see a bird in that situation, but I recognize an *invisible hand*, detaining the contented prisoner from her fields and groves for a purpose, as the event proves, the most worthy of the sacrifice, the most important, the most beneficial.[3]

If Darwin was arguing against Paley, and Paley was arguing against Smith, who was Smith arguing against? The mercantilists.

Throughout the early modern period—from the rise of the nation-state through the nineteenth century—the predominant economic ideology of the Western world was mercantilism, or the belief that nations compete for a fixed amount of wealth in a zero-sum game, born in part from our hunter-gatherer roots. Mercantilism assumes that the +X gain of one nation means the −X loss of another nation, with the +X and −X summing to zero. (All zero-sum games work in this fashion: if I beat you in a game by a score of 13–3, my net win was +10 and your net loss was −10, with the plus and minus figures canceling out to zero.) Therefore, in order for a nation to become wealthy, its government must run the economy from the top down through strict regulation of foreign and domestic trade, enforced monopolies, regulated trade guilds, subsidized colonies, accumulation of bullion and other precious metals, and countless other forms of economic intervention, all to the end of producing a "favorable balance of trade." Favorable, that is, for one nation over another nation.

Adam Smith was not an economist. He was, in essence, the titular founder of the science, and he studied what scholars of his age called "political economy." Smith was, in fact, a professor of moral philosophy at the University of Glasgow, where he taught courses in jurisprudence, ethics, rhetoric, and political economy. His first major work was *A Theory of Moral Sentiments*, published in 1759, in which he laid the foundation for the theory that we have an innate sense of morality: "How selfish soever man may be supposed, there are evidently some principles in his nature, which interest him in the fortune of others, and render their happiness necessary to him, though he derives nothing from it except the pleasure of seeing it. Of this kind is pity or compassion, the emotion which we feel for the misery of others, when we either see it, or are made to conceive it in a very lively manner." We sense someone else's joy or agony through empathy—by putting ourselves in their shoes and imagining how we would feel. "As we have no immediate experience of what other men feel," Smith writes, "we can form no idea of the manner in which they are affected, but by conceiving what we ourselves should feel in the like situation."[4] Out of our empathic nature comes the basis of a civil society—as we endeavor to assuage the distress we feel at others' distress, we attenuate our negative emotions and accentuate our positive passions:

> Upon these two different efforts, upon that of the spectator to enter into the sentiments of the person principally concerned, and upon that of the person principally concerned, to bring down his emotions to what the spectator can go along with, are founded two different sets of virtues. The soft, the gentle, the amiable virtues, the virtues of candid condescension and indulgent humanity, are founded upon the one: the great, the awful and respectable, the virtues of self-denial, of self-government, of that command of the passions which subjects all the movements of our nature to what

our own dignity and honour, and the propriety of our own conduct require, take their origin from the other.[5]

A *Theory of Moral Sentiments* was followed in 1776 with Smith's treatise on political economy, *An Inquiry into the Nature and Causes of the Wealth of Nations*. Smith's case against mercantilism is both moral and practical. It is moral, he argued, because "to prohibit a great people . . . from making all that they can of every part of their own produce, or from employing their stock and industry in the way that they judge most advantageous to themselves, is a manifest violation of the most sacred rights of mankind."[6] It is practical, he showed, because "whenever the law has attempted to regulate the wages of workmen, it has always been rather to lower them than to raise them."[7]

The *Wealth of Nations* was one long argument against the mercantilist system of protectionism and special privilege that in the short run may benefit producers but which in the long run harms consumers and thereby decreases the wealth of a nation. All such mercantilist practices benefit the producers and monopolists and their government agents, while the people of the nation—the true source of a nation's wealth—remain impoverished: "The wealth of a country consists, not of its gold and silver only, but in its lands, houses, and consumable goods of all different kinds." Yet "in the mercantile system, the interest of the consumer is almost always constantly sacrificed to that of the producer."[8]

The solution? Hands off. Laissez faire. Lift trade barriers and other restrictions on people's economic freedoms and allow them to exchange as they see fit for themselves, both morally and practically. In other words, an economy should be consumer driven, not producer driven. For example, under the mercantilist philosophy, cheaper foreign goods benefit consumers but they hurt domestic producers, so the government should impose protective trade tariffs to maintain the favorable balance of trade. But who is being protected by a protective tariff? Smith showed that, in principle, the mercantilist system only benefits a handful of producers while the great majority of consumers are further impoverished because they have to pay a higher price for foreign goods. The growing of grapes in France, Smith noted, is much cheaper and more efficient than in the colder climes of home, where "by means of glasses, hotbeds, and hotwalls, very good grapes can be raised in Scotland" but at a price thirty times greater than in France. "Would it be a reasonable law to prohibit the importation of all foreign wines, merely to encourage the making of claret and burgundy in Scotland?" Smith answered the question by invoking a deeper principle: "What is prudence in the conduct of every private family, can scarce be folly in that of a great kingdom. If a foreign country can supply us with a commodity cheaper than we ourselves can make it, better buy it of them."[9]

This is the core of Smith's economic theory: "Consumption is the sole end and purpose of all production; and the interest of the producer ought to be attended to, only so far as it may be necessary for promoting that of the consumer." The problem is that the system of mercantilism "seems to consider production, and not consumption, as the ultimate end and object of all industry and commerce."[10] And when production is the object, rather than consumption, producers appeal to top-down regulators instead of bottom-up consumers. Instead of consumers telling producers what they want to consume, government agents and politicians tell consumers what and how much they will consume, and at what price. Domestically, the government intervenes through *tax favors* for businesses (current estimates put the figure at about $750 billion annually for the United States), *tax subsidies* for corporations (about $125 billion allocated to Fortune 500 companies annually), *regulations* (to control prices, imports, exports, production, distribution, and sales), and *licensing* (to control wages and protect jobs).[11] Internationally, interventions come primarily through taxes of varying names, including "duties," "imposts," "excises," "tariffs," "protective tariffs," "import quotas," "export quotas," "most-favored nation agreements," "bilateral agreements," "multilateral agreements," and the like.

Such agreements are never between the consumers of two nations; they are between the politicians and the producers of the nations. Consumers have no say in the matter, with the exception of indirectly voting for the politicians who vote for or against such taxes and tariffs. And they all sum to the same effect: the replacement of free trade with "fair trade" (fair for producers, not consumers), an evolved version of the mercantilist "favorable balance of trade" (favorable for producers, not consumers). The true zero-sum game of mercantilism is one in which producers win by the reduction or elimination of competition from foreign producers, while consumers lose by having fewer products from which to choose, along with higher prices and often lower quality products. The net result is a decrease in the wealth of a nation.

✦ ✦ ✦

The founders of the United States and the framers of the Constitution were heavily influenced by the Enlightenment thinkers of England and the Continent, including Adam Smith. Nevertheless, it was not long after the founding of the country before our politicians began to shift the focus of the economy from consumption to production. In 1787, the United States Constitution was ratified; according to Article I, Section 8, "The Congress shall have the power to lay and collect taxes, duties, imposts, and excises to cover the debts of the United States." As an amusing exercise in bureaucratic wordplay, consider the common usages of these terms in the *Oxford English Dictionary*. *Tax*: "a compulsory contribution to the support of government"; *duty*: "a payment to the public revenue levied upon the import, export, manufacture, or sale of certain commodities"; *impost*: "a tax, duty, imposition levied on merchandise"; *excise*: "any toll or tax." (Note the oxymoronic phrase "compulsory contribution" in the first definition.) Thus an alternative Article I, Section 8 reads: "The Congress shall have the power to lay and collect taxes, taxes, taxes, and taxes to cover the debts of the United States."

In Europe, mercantilists dug in while political economists, armed with the intellectual weapons provided by Adam Smith, fought back, wielding the pen instead of the sword. The nineteenth-century French economist Frédéric Bastiat, for example, was one of the first political economists after Smith to show what happens when the market depends too heavily on top-down tinkering from the government. In his wickedly raffish *The Petition of the Candlemakers*, Bastiat satirizes special interest groups—in this case candlemakers—who petition the government for special favors:

> We are suffering from the ruinous competition of a foreign rival who apparently works under conditions so far superior to our own for the production of light, that he is flooding the domestic market with it at an incredibly low price. . . . This rival . . . is none other than the sun. . . . We ask you to be so good as to pass a law requiring the closing of all windows, dormers, skylights, inside and

outside shutters, curtains, casements, bull's-eyes, deadlights and blinds; in short, all openings, holes, chinks, and fissures.[12]

Bastiat also demonstrated the difference between what is seen and what is not seen when governments intervene in the marketplace. A public works bridge, such as the infamous Alaskan "bridge to nowhere," is seen by all, gloried in by its producers, and appreciated by its few users. What is not seen, however, are all the products that would have been produced or the services provided by the monies that were taxed out of private hands in order to finance the public project. "There is only one difference between a bad economist and a good one," Bastiat continued; "the bad economist confines himself to the *visible* effect; the good economist takes into account both the effect that can be seen and those effects that must be *foreseen*."[13] On the flip side, Bastiat wonders, "How is it that millions of people every day cooperate with millions of others to get the bagel to your corner coffee shop? There's no office or government agency or central hub where all the commerce originates. The web of connections holding the system together is unseen." The web of connections is the scaffolding that buttresses the market structure and keeps it from collapsing into chaos.[14]

Zero-sum mercantilist models hung on through the nineteenth and twentieth centuries, even in America. Since a permanent, peacetime income tax was not ratified until 1913, for most of the country's first century the practitioners of trade and commerce were compelled to contribute to the government through various other taxes. Foreign trade could not meet the growing debts of the United States, so in 1887 the government introduced the Interstate Commerce Commission to take advantage of the growing size and power of the railroads and to answer political pressure from farmers who felt powerless against them. The ICC was initially charged with regulating the services of railroads between states, but then expanded to oversee trucking companies, bus lines, freight carriers, water carriers, oil pipelines, transportation brokers, and other carriers of commerce.[15] Regardless of its intentions, the ICC's primary effect was to limit the freedom of markets that reached beyond the borders of the states of America, at a time when the country was earnestly attempting to forge a national identity that could overcome the sectional (or, one might say, tribal) divisions that had put it in jeopardy only decades earlier.

The ICC opened the way for the Sherman Antitrust Act of 1890, which declared, "Every contract, combination in the form of trust or otherwise, or conspiracy, in restraint of trade or commerce among the several States, or with foreign nations, is declared to be illegal. Every person who shall make any contract or engage in any combination or conspiracy hereby declared to be illegal shall be deemed guilty of a felony," resulting in a massive fine, jail, or both. When stripped of the obfuscatory language of bureaucratese, the Sherman Antitrust Act and the precedent-setting cases that have been decided in the courts in the century since it was passed allow the government to indict an individual or a company on one or more of four crimes: (1) *price gouging* (charging more than the competition); (2) *cutthroat competition* (charging less than the competition); (3) *price collusion* (charging the same as the competition), and (4) *monopoly* (having no competition).[16] This was Katy-bar-the-door for antibusiness legislators and their zero-sum mercantilist bureaucrats to restrict the freedom of consumers and producers to buy and sell, which they did with reckless abandon. In the century since it was passed, no entity has received worse treatment than the Aluminum Company of America.[17]

In 1886, Charles Martin Hall solved the problem of how to commercially produce aluminum by a process of passing an electric current through a bath of cryolite and aluminum oxide that left a semirare metal byproduct: aluminum. With financial backing, Hall founded the Pittsburgh Reduction Company, which in 1907 changed its name to the Aluminum Company of America, or Alcoa. Shortly after Hall introduced his revolutionary new smelting system he managed to reduce the price of aluminum from $545 a pound to $8 a pound—a 98.5 percent cut—initially producing 10 pounds a day for a gross take of $80 a day. By the 1930s, the company was generating a million dollars a day, but instead of raising the price (as monopolies are alleged to do, following the folk intuitions that rule our economic understanding), Alcoa cut the price of aluminum to a mere 20 cents a pound, 97.5 percent off the existing price. As reward for this remarkable contribution to the wealth of the nation and its millions of consumers, the Department of Justice sued Alcoa in 1937, charging the company's directors on

140 criminal counts, including excessive prices! The trial lasted from 1938 to 1940, the longest trial in American history to that date. The transcript ran to over fifty thousand pages set in 480 volumes and weighing in at 325 pounds. The court ruled in favor of Alcoa, so the Justice Department appealed. On June 12, 1944, the decision was reversed and Alcoa lost on a single count of the original 140 charges—monopoly—and was forced to enable its competitors—Reynolds Metals Company and Kaiser Chemicals—to become fully competitive producers of primary aluminum. Judge Learned Hand's assessment of Alcoa's crimes is a model of double-think ratiocination:

> It was not inevitable that it should always anticipate increases in the demand for ingot and be prepared to supply them. Nothing compelled it to keep doubling and redoubling its capacity before others entered the field. It insists that it never excluded competitors; but we can think of no more effective exclusion than progressively to embrace each new opportunity as it opened, and to face every newcomer with new capacity already geared into a great organization, having the advantage of experience, trade connections and the elite of personnel.[18]

Stripping away the obscure bureaucratic legalese, we find that Alcoa's crimes included: (1) Alcoa anticipated consumer demands for its product and was prepared to supply them, (2) Alcoa doubled and redoubled its production capacity in order to meet the needs of its customers, (3) Alcoa embraced new opportunities, and (4) Alcoa had experience, connections, and trained employees. Was this hurting the millions of consumers of aluminum? No. It was hurting other producers of aluminum! In the name of protecting consumers from the alleged malfeasance of a monopoly, the U.S. Department of Justice protected a couple of aluminum producers at the expense of millions of consumers. Even if Alcoa was a monopoly (which for a time it was), and even if it contrived to raise, lower, or hold steady its prices in response to competitors, the point of this history is to note the effects of top-down intervention on the marketplace. Even granting that large corporate trusts occasionally do conspire to conduct business in a manner more conducive to producers (themselves) than consumers (their customers), I am interested here in weighing the long-term effects of Smith's deeper principle, that economies are best structured from the bottom up (consumer driven) and not from the top down (producer driven).[19]

Although the Alcoa story will sound antiquated to modern ears, what happened to that corporation is no different in principle from what happened to Microsoft when it was similarly charged in the 1990s as allegedly in violation of the Sherman Antitrust Act. Microsoft was accused of gaining a market advantage over its competitors through the wildly successful Windows operating system by adding to it a free version of a Web browser, Internet Explorer, which competed with other browsers such as Netscape's; these other browsers were not free. Microsoft's crime was to offer special discounts to major vendors such as IBM, Intel, and Compaq as an incentive to adopt Microsoft technology. One of these vendors was America Online (AOL), for which Microsoft developed a browser designed specifically for its Internet service. In exchange for AOL adopting Microsoft's Internet software, Microsoft provided AOL with free worldwide distribution rights to Internet Explorer and placement of the AOL icon in a special folder on the Windows desktop. The effects were immediate and dramatic. AOL quickly registered nearly one million new subscribers, and soon tens of millions of Internet consumers could access cyberspace at no additional cost. Microsoft offered Internet Explorer free to consumers. Surely this is a good thing, no?

Not according to the Department of Justice, which charged Microsoft with monopolistic practices. Here is what United States District Court judge Thomas Penfield Jackson had to say about Microsoft and its evil doings in his judgment against them on November 5, 1999:

> The inclusion of Internet Explorer with Windows at no separate charge increased general familiarity with the Internet and reduced the cost to the public of gaining access to it, at least in part because it compelled Netscape to stop charging for Navigator. These actions thus contributed to improving the quality of Web browsing software, lowering its cost, and increasing its availability, thereby benefiting consumers.[20]

This is a crime? Yes, because "Microsoft also engaged in a concerted series of actions designed to protect the applications barrier to entry, and hence its monopoly power, from a variety of middleware threats, including Netscape's Web browser and Sun's implementation of Java."[21] According to the judge, "This indicates that superior quality was not responsible for the dramatic rise in Internet Explorer's usage share."[22] In other words, even though Microsoft offered a higher quality product at a lower price, that is not what led to its success over Netscape; rather, Microsoft's exclusive deals and special offers to other companies with whom it desired to do business is what led to its success, and that is not fair. Not fair to whom? Consumers? No—as Judge Jackson admitted, it was a boon to consumers. So to whom was it not fair? The answer should be obvious by now: other producers.

✦ ✦ ✦

To those who fear monopolies and believe that some producers should be protected from other producers by the government, such antitrust suits can be considered morally necessary. But the issue is not whether antitrust legislation is moral (although I do believe it is immoral); it is whether it serves the free market and our folk economic intuitions. Antitrust actions are premised on a win-lose, zero-sum, producer-driven economy, but the economy is not zero sum. And it is our folk intuitions that lead us to believe that the economy must have been designed from the top down, and thus can only succeed with continual tinkering and control from the top. But in his evolutionary, bottom-up model of the economy, Smith amassed copious evidence that counters this myth to show that, in the modern language of complexity theory, the economy is a self-organized emergent property of complex adaptive systems.

Adam Smith launched a revolution that has yet to be fully realized. A week does not go by without a politician, economist, or social commentator bemoaning the loss of American jobs, American manufacturing, and American products to foreign jobs, foreign manufacturing, and foreign products. Even conservatives—allegedly in favor of free markets, open competition, and less government intervention in the economy—have few qualms about employing protectionism when it comes to domestic producers, even at the cost of harming domestic consumers. Even that icon of free market capitalism, President Ronald Reagan, compromised his principles in 1982 to protect the Harley-Davidson Motor Company when it was struggling to compete against Japanese motorcycle manufacturers that were producing higher-quality bikes at lower prices. Honda, Kawasaki, Yamaha, and Suzuki were routinely undercutting Harley-Davidson by $1,500 to $2,000 a bike in comparable models. A true defender of free market evolutionary economics would have cheered this spectacular savings to American consumers. After all, what difference does it make to consumers who produces the products that they want? But on January 19, 1983, the International Trade Commission ruled that foreign motorcycle imports were a threat to domestic motorcycle manufacturers, and a 2-to-1 finding of injury was ruled on the petition by Harley-Davidson, which complained that it could not compete with foreign motorcycle producers.[23] On April 1, Reagan approved the ITC recommendation, explaining to Congress, "I have determined that import relief in this case is consistent with our national economic interest," thereby raising the current tariff of 4.4 percent to 49.4 percent for a year, a tenfold tax increase on foreign motorcycles that would have to be absorbed by American consumers. The protective tariff worked to help Harley-Davidson recover financially, but it was American motorcycle consumers who paid the price, not Japanese producers. As ITC chairman Alfred E. Eckes explained about his decision: "In the short run, price increases may have some adverse impact on consumers, but the domestic industry's adjustment will have a positive long-term effect. The proposed relief will save domestic jobs and lead to increased domestic production of competitive motorcycles."[24]

Whenever free trade agreements are proposed that would allow domestic manufacturers to produce their goods cheaper overseas and thereby sell them domestically at a much lower price than they could have with domestic labor (and thus benefit domestic consumers and the overall wealth of the nation), politicians and economists, under pressure from trade unions and political constituents, routinely respond disapprovingly, claiming that we must protect our workers. Recall presidential candidate Ross Perot's oft-quoted 1992 comment in response to the North American Free Trade Agreement (NAFTA) about the "giant sucking sound" of jobs being sent to Mexico from the United States. Here is the protectionist formula: domestic workers win while foreign workers lose and domestic consumers lose. This win-lose-lose game continues to be played today.

If Bill Gates is the Antichrist in the left's pantheon of fallen gods, for those who embrace producer-directed economics, Wal-Mart resides just below the counterfeiters and falsifiers in the eighth circle of Dante's inferno. By dramatically slashing prices and undercutting the competition of smaller retail chains and mom-and-pop stores—causing many of them to go out of business—Wal-Mart has hurt the American economy, right? Wrong. By employing 1.3 million people (about as many as the military), and keeping retail prices low through quantity purchasing, a McKinsey & Company study estimated that Wal-Mart alone accounted for a whopping 13 percent of U.S. productivity gains in the second half of the 1990s. As the savvy social commentator and political analyst George Will noted, for every fifty retail jobs that Wal-Mart caused to be lost among its competitors, it created a hundred new jobs at Wal-Mart, making it "about as important as the Federal Reserve in holding down inflation."

Nevertheless, during the 2004 presidential campaign, John Kerry called Wal-Mart "disgraceful" and emblematic of "what's wrong with America." That's from a producer-directed perspective, of course. From a consumer-directed perspective, consumers consume Wal-Mart products because they can save money doing so, and workers work at Wal-Mart because they perceive it to be in their best interests. As Will notes, Wal-Mart prices average 17 percent less than those of other retail outlets, drawing in 127 million customers per week, and in 2006 more than twenty-five thousand people applied for 325 jobs at a Wal-Mart in Evergreen Park, Illinois. Calling Wal-Mart "the most prodigious job-creator in the history of the private sector in this galaxy," Will estimates that "Wal-Mart and its effects save shoppers more than $200 billion a year, dwarfing such government programs as food stamps ($28.6 billion) and the earned-income tax credit ($34.6 billion)."[25] To paraphrase Yogi Berra, if people won't work and shop at other retail stores, there's no way you can stop them.

In early 2007, the Nobel laureate economist Edward C. Prescott lamented that economists invest copious time and resources countering the myth that it is "the government's economic responsibility to protect U.S. industry, employment and wealth against the forces of foreign competition." That is not the government's responsibility, says Prescott, echoing Smith, maintaining instead that it is simply "to provide the opportunity for people to seek their livelihood on their own terms, in open international markets, with as little interference from government as possible." Prescott shows that "those countries that open their borders to international competition are those countries with the highest per capita income" and that open economic borders are "the key to bringing developing nations up to the standard of living enjoyed by citizens of wealthier countries." Noting that 2007 marks the fiftieth anniversary of the signing of the Treaty of Rome, in which France, Italy, Belgium, West Germany, Luxembourg, and the Netherlands formed what would eventually become the European Union by opening their borders to free trade, they moved from being half as productive as the United States to being on an equal footing in only half that time. By comparison, Denmark, Ireland, and the United Kingdom were on economic par with those six countries before they (the six) signed the treaty, but they subsequently fell behind the six. When Denmark, Ireland, and the United Kingdom opened their economic borders to trade with the original Treaty of Rome six, they eventually caught up, and the United Kingdom is today as productive as Germany.[26]

Such examples do not represent controlled experiments, but we can employ an additional test, in which we compare social outcomes where influencing factors are varied by natural means. Employing this comparison method lends additional support for the power of free markets and open competition. In the 1980s, Spain, Portugal, and Greece joined the free market club, quickly bringing Spain up to equal footing with the original Rome six, with Portugal and Greece not far behind. Austria, Sweden, and Finland joined in 1995, and their economic fortunes have taken an upward turn. Ten more countries have joined since 2004, and they are beginning to show signs of upward economic mobility. Similarly, the five wealthiest East Asian countries of Taiwan, Singapore, Japan, South Korea, and Hong Kong experienced upturns in their economies once they loosened the chains that bound their markets. By comparison, Prescott presents research showing the effects of protectionism on the wealth of a nation. From 1950 to 2001, for example, the per capita GDP for Europe increased 68 percent compared to that of the United States, and Asia's increased 244 percent relative to the United States; by comparison, those

Latin American countries closed to international trade saw a relative *decrease* of 21 percent, whereas in 1950 those same Latin American countries had exceeded Asia's GDP by 75 percent.[27]

The comparison method can also be employed to an unfolding economic experiment in the form of the Central American Free Trade Agreement (CAFTA). In the year since CAFTA was signed in March 2006 and the United States began trading with El Salvador, Guatemala, Nicaragua, and Honduras, imports from America to these Central America countries increased significantly, along with investment capital into key industries. In the second half of 2006, for example, after it joined CAFTA, total trade in Guatemala grew by 17 percent, compared with a 5 percent growth rate in the first half of the year. Nicaragua is experiencing similar economic improvement. Between April 2006, when it signed on to CAFTA, and the end of that year, total trade increased by more than 20 percent compared to the same period in 2005.[28]

This is a brainful of data, but these and other natural experiments from around the world allow us to empirically test the cause-and-effect relationships between free trade and economic prosperity. "Protectionism is seductive," Prescott admits, "but countries that succumb to its allure will soon have their economic hearts broken. Conversely, countries that commit to competitive borders will ensure a brighter economic future for their citizens." Why do open economic borders, free trade, and international competition lead to greater wealth for a nation? Writing more than two centuries after Adam Smith, Prescott reverberates the moral philosopher's original insight: "It is openness that gives people the opportunity to use their entrepreneurial talents to create social surplus, rather than using those talents to protect what they already have. Social surplus begets growth, which begets social surplus, and so on. People in all countries are motivated to improve their condition, and all countries have their share of talented risk-takers, but without the promise that a competitive system brings, that motivation and those talents will only lie dormant."[29]

✦ ✦ ✦

Why is mercantilist zero-sum protectionism so pervasive and persistent? I have shown that bottom-up invisible hand explanations for complex systems are counterintuitive because of our folk economic propensity to perceive designed systems to be the product of a top-down designer. But there is a deeper reason, grounded in our evolved social psychology of group loyalty: our propensity to circle the wagons and protect our own.

Only in the last ten thousand years have bands of hundreds evolved into tribes of thousands, with some tribes developing into chiefdoms of tens of thousands, and some chiefdoms coalescing into states of hundreds of thousands, and a handful of states conjoining into empires of millions. The attendant leap in food production and population that accompanied the shift to chiefdoms and states allowed for a division of labor to develop in both the economic and social spheres. Full-time artisans, craftsmen, and scribes worked within a social structure organized and run by full-time politicians, bureaucrats, and, to pay for it all, tax collectors. The modern state economy was born.

In this historical trajectory our group psychology evolved, and along with it a propensity for xenophobia—in-group good, out-group bad. In the Paleolithic social environment in which our moral commitments evolved, one's fellow in-group members consisted of family, extended family, friends, and community members who were well known to each other. To help others was to help oneself. Those groups who practiced in-group harmony and between-group antagonism would have had a survival advantage over those groups who experienced within-group social divide and incoherence, or haphazardly embraced strangers from other groups without first establishing trust. Because our deep social commitments evolved as part of our behavioral repertoire of responses for survival in a complex social environment, we carry the seeds of such in-group inclusiveness today. The resulting within-group cohesiveness carries with it a concomitant tendency for between-group xenophobia and tribalism that, in the context of a modern economic system, leads to protectionism and mercantilism.

◆ ◆ ◆

Adam Smith was anything but blindly probusiness. In fact, Smith was quite skeptical of the base motives of producers. Throughout *The Wealth of Nations*, Smith criticized "factions"—groups of politically connected businessmen, bankers, tradesmen, and industrialists who turned to the government to do their bidding—because their formation constituted a power bloc that would serve the special interests of producers rather than the general interests of consumers: "People of the same trade seldom meet together, even for merriment and diversion, but the conversation ends in a conspiracy against the public, or in some contrivance to raise prices."[30] He also did not argue that self-interest is always good. He believed that self-interest is a part of human nature (along with empathy and other prosocial sentiments) and is therefore not necessarily bad. But he never pulled his punches against those who he perceived acted out of excessive greed and avarice, observing, "We are not ready to suspect any person of being defective in selfishness," and warning that "all for ourselves, and nothing for other people, seems, in every age of the world, to have been the vile maxim of the masters of mankind."[31]

We have both virtues and vices, which interact and feed on one another, and whose relative expression depends on the social context. Mutually beneficial free-exchange environments produce cooperativeness even while each party may be motivated by competitiveness. As long as there is no fraud or coercion in the exchange, buyers and sellers both win when they agree upon the reciprocally advantageous terms of the trade. The buyer selfishly values the seller's product more than he does his money, and the seller greedily values the buyer's money more than he does his product. As Smith explained in one of the most famous passages in *The Wealth of Nations*: "It is not from the benevolence of the butcher, the brewer, or the baker that we expect our dinner, but from their regard to their own interest. We address ourselves, not to their humanity but to their self-love, and never talk to them of our own necessities but of their advantages."[32]

By allowing people to follow their natural inclination to pursue their self-love, the country as a whole will prosper, almost as if the entire system were being directed by a magical force. It is here that we find the one and only use in *The Wealth of Nations* of the most famous metaphor in Western economic thought:

> Every individual is continually exerting himself to find out the most advantageous employment for whatever capital he can command. . . . He generally, indeed, neither intends to promote the public interest, nor knows how much he is promoting it. He

intends only his own security; and by directing that industry in such a manner as its produce may be of the greatest value, he intends only his own gain, and he is in this, as in many other cases, led by an *invisible hand* to promote an end which was no part of his intention. By pursuing his own interest he frequently promotes that of the society more effectually than when he really intends to promote it.[33]

The economy is a product of human action, not of human design. A system driven by an invisible hand is *decentralized*, and therefore complex organization within the system is an unintended *by-product* of the baser instincts of the economic agents. This brings us back full circle to Darwin and what happens in nature when organisms pursue their self-love, with no cognizance of the unintended consequences of their behavior:

> It may be said that *natural selection* is daily and hourly scrutinising, throughout the world, every variation, even the slightest; rejecting that which is bad, preserving and adding up all that is good; silently and insensibly working, whenever and wherever opportunity offers, at the improvement of each organic being in relation to its organic and inorganic conditions of life. We see nothing of these slow changes in progress, until the hand of time has marked the long lapses of ages, and then so imperfect is our view into long past geological ages, that we only see that the forms of life are now different from what they formerly were.[34]

These descriptors—*invisible hand* and *natural selection*—are so powerful, and so deeply annealed into our psyches, that it is difficult not to think of them as *forces* of nature, such as gravity and electromagnetism, or as mechanical systems, such as gears and pulleys. But they are not forces or mechanisms, because there is nothing acting on the agents in the system in such a causal manner. Instead, Smith's invisible hand and Darwin's natural selection are *descriptions* of processes that naturally occur in the economies of nature and society. The causal mechanisms behind the invisible hand and natural selection lie elsewhere in the system—within the agents themselves—which is why Smith invested so much work on understanding the natural sympathies of people and Darwin advanced so much effort toward comprehending the natural tendencies of organisms.[35]

Adam Smith showed how national wealth and social harmony were unintended consequences of individual competition among people. Charles Darwin showed how complex design and ecological balance were unintended consequences of individual competition among organisms. The artificial economy mirrors the natural economy.[36]

◆　◆　◆

If economies are best driven from the bottom up and not the top down, and if consumer-driven free markets are more fair and efficient than producer-driven mercantilist markets, how far can we take this generalization? There are extremists who embrace the ideology of *anarcho-capitalism*, which holds that political systems will eventually fall into disuse (in the same manner but for different reasons and with different results than Karl Marx predicted). Public social services will be privatized gradually so that eventually the entire planet will be one global free market with no economic or political borders, with no hostile states, with minimum crime and criminals, and in which disputes will be settled through private arbitration. Transhumanists—those who believe that one day we will evolve into genetically engineered biorobots capable of significantly greater strength, intelligence, and lifespan—even envision the day when this free market society will not only span the globe, but extend out into space as we colonize a terraformed Mars. Eventually transhumans will establish societies on the moons of Jupiter and Saturn and in other solar systems, and ultimately, over the course of millions of years, the galaxy itself will become an integrated market of explorer-colonizer-traders. Who knows, some futurists imagine—perhaps a billion years from now we will even be colonizing other galaxies, and in a few tens of billions of years the entire cosmos.

Such utopian visions sound so Lennonesque! Imagine no borders, no countries. No need for greed or hunger. A brotherhood of man. But as the sage pop philosopher Yogi Berra once said, "In theory, there is no difference between theory and practice. In practice, there is."

In order to keep the free market both free and fair, we need political states based on the rule of law, with property rights, a secure and trustworthy banking and monetary system, economic stability, a reliable infrastructure, protection of civil liberties, a clean and safe environment, and various freedoms: of movement, of the press, of association, and of education. We need a robust military for protection of our liberties from attacks by other states. We need a potent police force for protection of our freedoms from attacks by other people within the state. We need a viable legislative system for establishing fair and just laws. We need an effective judicial system for the equitable enforcement of those fair and just laws.

The best politico-economic system to date is a liberal democracy and free market capitalism, or *democratic-capitalism*. In a system of democratic-capitalism, social liberalism and fiscal conservatism is a synergistic marriage that leads to the greatest prosperity, the greatest liberty, and the greatest happiness for the greatest number.

4

OF PANDAS, PRODUCTS, AND PEOPLE

In 1979, I took up the sport of bicycle racing, and by the next year I was serious enough about competing that I purchased the top-of-the-line Bianchi racing bike—an Italian import whose frame was constructed of the super-lightweight but ultrastrong Reynolds 531 double-butted steel tubing painted in Bianchi's classic celeste green, equipped with Campagnolo's Super Record components (parts of which were drilled out to reduce weight), and featuring a Brooks leather saddle with brass rivets, hand-sewn leather handlebar wraps, ultralight sew-up tires, and titanium toe clips and freewheel cluster for additional weight diminution. It was a dream machine that glided down the road in effortless silence.

I never imagined how that bike could be improved, or how and why the market trends at that time would ever shift. For serious cycling, Reynolds (and Columbus) dominated the tubing market, Campagnolo owned the component market, and European imports controlled the brand preference market. A bike frame with the unmistakable TI Reynolds light green sticker with the metallic gold lettering was a sign of excellence. "Campy" was considered the crème de la crème of components ever since the Italian bike racer Tullio Campagnolo converted his frustration with failed equipment during a bike race into a manufacturing plant that came to symbolize professionalism. Serious cyclists would no more ride without Campy on their frames than they would race without shaved legs. It had been this way for decades, and everyone knows that historical momentum makes markets resistant to change, right?

Wrong. By the mid-1980s, we were racing on frames made of aluminum, titanium, carbon fiber, or some combination thereof; pedals with toe clips had been almost entirely displaced by click-in clipless pedals introduced by the manufacturer of ski bindings; heavy leather saddles were replaced by lighter models made of composite materials and covered in comfortable padding; and our beloved Campy barely survived a marketing onslaught from the Japanese upstart Shimano, who gave us an assembly of technological improvements in component design and performance that far surpassed anything Campy offered, most notably the Dura-Ace Shimano Index Shifting (SIS) system that perfectly locked in each gear shift and moved the shifters from the down-tube to the brake hoods (thereby eliminating the necessity of taking your hands off the bars). Shimano turned 10-speeds into 12-, 14-, 16-, even 18-speeds. What seemed at the time to be a market locked into its path by historical momentum and economic dominance by a few companies was far more vulnerable to buffeting from outside forces and innovative competition than anyone imagined.

◆　◆　◆

Are markets determined by a series of quirky, chancy, contingent events—a chain of accidents? Or are markets shaped by the ineluctable forces of the physical, economic, and social worlds—a rule of law? If we rewound the tape of economic history and played it back, would we end up with a set of markets, industries, businesses, and products similar to what we have today?

This turns out to be a deep question with profound implications for matters well outside the purview of economics. For example, was our very existence foreordained from the beginning of the universe because of the particular configuration of the laws of nature, or are we nothing more than a fluke of history, the end product of a near-endless chain of accidents? The great debate here is between what happens necessarily and what happens contingently—what *must* be versus what *may* be. Put more formally, is the current state of things a *necessity*—it could not have been otherwise?—or is it a *contingency*—it need not have been?

This is an ancient question dating back to Aristotle, but the most recent rendition of it was popularized by Stephen Jay Gould, most notably in his 1989 book, *Wonderful Life*. Gould suggested that if the tape of life were rewound 530 million years to the time of the organisms found in the Canadian outcrop known as the Burgess Shale and replayed with a few contingencies tweaked here and there, humans would most likely never have evolved. The eventual rise of humans includes millions of antecedent states in our past. Each event in the sequence has a cause, and thus is determined, but the eventual outcome is unpredictable because of the complex array of contingencies involved. So powerful are the effects of contingency, Gould believed, that a small change in the early stages of a sequence produces large effects in the later stages:

> I am not speaking of randomness (for E had to arise, as a consequence of A through D), but of the central principle of all history—contingency. A historical explanation does not rest on direct deductions from laws of nature, but on an unpredictable sequence of antecedent states, where any major change in any step of the sequence would have altered the final result. This final result is therefore dependent, or contingent, upon everything that came before—the unerasable and determining signature of history.[1]

The power of contingency was also well captured by the physicist Edward Lorenz in his famous paper "Does the Flap of a Butterfly's Wings in Brazil Set Off a Tornado in Texas?"[2] This "butterfly effect" means that the scientist's search for general laws and universal principles must be tempered by the capricious happenstances of history.

Gould's type specimen of contingency is the panda's thumb. In a 1978 essay, "The Panda's Peculiar Thumb," Gould showed that the panda's thumb is not a predictable design of nature's necessitating laws of form, but an improvised contraption constructed from its evolutionary history.[3] The panda's thumb, in fact, is an example of a bottom-up design, the product of the tinkering of evolution, which uses whatever biological equipment is available. A top-down intelligent designer would surely have created a much more elegant and efficient thumb for the panda to strip the leaves off bamboo shoots. Instead, its feeble little nub of a digit is not a thumb at all. The digit it uses for leaf harvesting is called its "thumb," but it is actually an extended radial sesamoid (wrist) bone. The panda's paw already has the traditional five digits all locked into place through muscles, tendons, and nerves designed by evolution for clawing from forward to back, as do all bears and other mammalian carnivora. In other words, the panda's extra digit is not an intelligently designed thumb, but constructed from the available parts bestowed upon it by its history; that is, it is a jury-rigged wrist bone. Compared to our thumb, the panda's thumb may appear clumsy and awkward, but it is good enough for reaping its plant rewards and thus there was no reason for natural selection to undertake a major overhaul of the entire paw.

✦ ✦ ✦

In economics, this evolutionary process is called *path dependency*, or *historical lock-in*, where markets become dependent on the paths they are already on, or they become locked into the channels in which they are operating. The concept of path dependency was introduced by the Stanford University economist Paul David, whose seminal 1985 paper, "Clio and the Economics of QWERTY," launched the keyboard legend.[4] Technically, David defined path dependency as a "property of contingent, non-reversible dynamic processes, including a wide array of processes that can properly be described as 'evolutionary.'" Or, more succinctly, *history matters.*[5]

One example of path dependency is the Bestseller Effect discussed in the prologue, by which the rich get richer. More broadly, path dependency predicts that companies who get into an industry early have a head start on later competitors, gaining a competitive edge through a number of factors. In production, for example, economies of scale mean that fixed costs can be distributed throughout the production process so that the more you produce, the more you can lower the cost of the finished product to consumers, thereby capturing a larger share of the market. First, this creates a *bandwagon effect*, causing consumers (with the exception of early adopters) to gravitate toward products that they think will most likely become readily available, as assessed by price and convenience of purchase. Second, this creates a *network effect*, when producers and retailers, anticipating the bandwagon effect, produce and stock up on the products they think will be most in demand by consumers. A feedback loop is then established whereby the most popular products are produced in the greatest quantity, reinforcing consumers to consume more of them, leading to more production, and so on.

More consumers using a product also locks them into a pattern of consumption that may preclude using other technologies, either because there is a physical incompatibility or because there is a prohibitive switching cost. If Microsoft captures a significant percentage of the market of users of computer operating systems with Windows, and Windows precludes the use of other operating systems, then consumers locked into Windows are less likely to hassle switching to a new system. Once a social network of consumers is captured by one producer, it means that even if a higher quality product at a lower price is offered, it will struggle to survive because the entire market has become dependent on the path that was dug by those early entrants into the industry. The market, at that point, has reached a temporary state of equilibrium, optimization, or stability.

Technically speaking, staying with the old technology may sometimes be what is called a *Nash equilibrium*, a concept identified by the Nobel laureate mathematician John Nash (immortalized in the film *A Beautiful Mind*), in which two or more players reach an equilibrium where neither one has anything to gain by unilaterally changing strategies. If each player has selected a tactic such that no player can benefit by changing tactics while the other players hold to their plans, then that particular arrangement of strategy choices is said to have reached a point of equilibrium.[6] Applied to economics, markets reach a point of equilibrium where holding on to strategies is more profitable (or at least is perceived to be so) than switching strategies; thus, industries, corporations, businesses, and people reach a point of stability in which switching strategies appears undesirable.

In a related concept more directly applied to economic models, such product switching may not be *Pareto efficient*, a concept derived by the

Italian economist Vilfredo Pareto from his studies of market efficiency. The question on the table was this: are free and competitive markets efficient in terms of optimizing the highest-quality products at the lowest prices? To answer the question, Pareto identified four types of trades made by people in markets: *win–win* (all parties gain), *win–no-lose* (some parties gain while other parties neither gain nor lose), *no-lose–lose* (no one gains but someone loses), and *win–lose* (some gain while others lose). Pareto's theory holds that in the long run, parties would only continually engage in trades in which either both parties gain or one party gains but the other does not lose; that is, either *win–win* or *win–no-lose* trades. Eventually markets would reach a point of equilibrium, where an optimum level of *win–win* and *win–no-lose* trades is reached and at which no further trades could be made without someone losing. If we define economics as the allocation of scarce resources that have alternative uses, given a set of individuals and a set of allocations from which to choose, switching from one allocation to another that makes at least one individual better off without making anyone else worse off is a state of *Pareto optimization*. Where no further improvement can be made, the allocation of resources is *Pareto efficient*.[7]

In evolutionary theory, such equilibriums, efficiencies, and optimizations are similar to a concept known as *Evolutionary Stable Strategies*, or a strategy that when adopted by a population of individuals consistently outcompetes alternative strategies.[8] In the panda example, the strategy of stripping leaves off bamboo shoots with the modified wrist bone was selected for because even though it is a suboptimal design, it is still more efficient than redesigning the entire paw to evolve a thumb. The panda's thumb is an Evolutionary Stable Strategy. So too is the QWERTY keyboard.

In 1873, Christopher Latham Sholes sold the rights to his typewriter patent—that included the QWERTY keyboard—to E. Remington & Sons. In 1882, the Shorthand and Typewriter Institute in Cincinnati was founded by L. V. Longley, who chose to adopt, from among many keyboard arrangements, the QWERTY system. As her typing program became more successful, her teaching methods were adopted by others until they became the industry standard, even adopted by Remington when it began to establish typing schools. Then, in 1888, a typing contest was arranged between Longley's method and that of her competitor Louis Taub, who employed a different typing technique on a non-QWERTY keyboard. Longley's star pupil, Frank McGurrin, purportedly thrashed the competition by memorizing the QWERTY keyboard and employing the newfangled method of touch typing. The event generated much publicity, and touch typing became the method of choice for American typists. The QWERTY keyboard on the Remington typewriter became necessary to the point that it would take a typing revolution to reshuffle the keyboard deck.[9] So says the legend, which has since been printed as fact.

According to the myth, QWERTY was designed for nineteenth-century typewriters whose key striking mechanisms were too slow for human finger speed, which at full finger function would cause keys to constantly jam. Even though more than 70 percent of English words can be produced with the letters DHIATENSOR, a quick glance at the keyboard will show that most of these letters are not in a strong striking position (home row struck by the strong first two fingers of each hand). All the vowels in QWERTY, in fact, are removed from the strongest striking positions, leaving only 32 percent of the typing on the home row. Only about a hundred words can be typed exclusively on the home row, while the weaker left hand is required to type over three thousand different words alone, not using the right hand at all. Another check of the keyboard reveals the alphabetic sequence (minus the vowels) DFGHJKL. It appears that the original key arrangement was just a straight alphabetical sequence, which makes sense in early experiments before testing was done to determine a faster alignment. The vowels were removed to slow the typist and prevent key jamming. This problem was eventually remedied, but by then QWERTY was so entrenched in the system (through manuals, teaching techniques, and other social necessities) that unless the major computer companies, along with keyboard teachers, manuals, and a majority of typists all decided to change simultaneously, we are stuck with QWERTY indefinitely.

✦ ✦ ✦

The story of the origin and persistence of the QWERTY keyboard as an example of path dependency is so compelling that it is told and retold by numerous authors. Picking up on David's scholarly articles and Gould's popular works, economists, social commentators, and journalists repeat the legend as if it were an economic law that history trumps markets, and that market success is determined more by the luck of the draw than the quality of the product. But if path dependency is such a potent force in economics, why are there not dozens or hundreds of examples readily available? The answer seems to be that the QWERTY story is largely mythical. Although it is true that QWERTY came to us through a series of contingent and quirky events—and thus, in this sense, history matters—it turns out to be anything but suboptimal. It appears that QWERTY is as fast and efficient as any keyboard alternative, and thus is more analogous to the human effectual thumb than the panda's peculiar thumb.

As a test of QWERTY path dependency, comparisons are often made to the Dvorak Simplified Keyboard (DSK) patented in 1936 by August Dvorak, which is claimed to be more efficient than QWERTY's allegedly suboptimal layout. If Dvorak is so superior, then why isn't there a market for it? The answer commonly given is the path dependency of QWERTY momentum—there are too few people trained to use the Dvorak keyboard and thus there is no demand for keyboard manufacturers to meet for this superior design.

This explanation, however, has been challenged. New historical research shows that Sholes originally designed QWERTY to prevent key jamming, not by slowing down the typist, but by separating keys whose type-bar letters were close to each other underneath the typewriter carriage. (He moved the T and the H apart, for example.) The QWERTY design was really a result of Sholes's study of letter pair frequencies, not an impediment to fast typists whose alacrity conflicted with the mechanical ability of typewriters of that time.

Indeed, economists Stan Liebowitz and Stephen Margolis tracked down the test results that supposedly proved the Dvorak keyboard superior to QWERTY, and their account can best be described as less than charitable for Dvorak.[10] For example, in one study, Dvorak compared students of different ages, from different schools, in classes with different lengths of training time, taking different typing tests. "One doesn't need to be a scientist to realize that such comparisons are not the stuff of controlled experiments," Liebowitz and Margolis sardonically note.[11] In addition, the results of an oft-cited 1944 navy test that forms the entire foundation for the claim that Dvorak is superior to QWERTY appear to be cooked. For starters, Liebowitz and Margolis initially could not find anyone who had a copy of the navy study, concluding that "this appears to be one of those cases in which one author relies on another's account, who in turn is relying on another's and so on, without any of them reading the original."[12] When they finally tracked down a copy in the attic of a farmhouse in Vermont, the headquarters of a group calling itself Dvorak International, they discovered that the study was conducted by none other than navy lieutenant August Dvorak himself, making his data rather suspect. The study was fraught with methodological problems. Navy typists received different amounts of training time for QWERTY versus Dvorak, and their initial typing scores were measured differently across the two groups. Even worse, it appears that three typists in the QWERTY group had initial net scores of zero words per minute—and these null results were included in the calculation of the QWERTY average typing speeds, which Liebowitz and Margolis compute halved QWERTY's average. The test was bogus.

The fact is QWERTY is good enough to get the job done and has consistently met all would-be competitors, not with unfair market momentum, but with equal or superior performance. It turns out that there were lots of typing competitions in the late nineteenth century, with many different keyboard arrangements and with plenty of people who could touch-type, and still QWERTY held its own on performance criteria alone. Liebowitz and Margolis found a contest reported in the *New York Times* that was held in August 1888—just weeks after McGurrin's Cincinnati victory—in which a competitor named May Orr typed 95.2 words per minute while another competitor named M. Grant typed 93.8 words per minute. Also in the contest: Frank McGurrin, who won by a slim margin at 95.8 words per minute.

History matters, but so too does quality. QWERTY may be suboptimal, but it is no less so than its erstwhile competitors. QWERTY may not be perfect, but it is good enough to get the job done.

◆ ◆ ◆

The deeper implication of the theory of path dependency is that when markets are left alone to evolve from the bottom up without top-down controls, they are inefficient and do not deliver the best products. In a 1990 article in *Scientific American*, the economist Brian Arthur noted the deleterious effects of path dependency: "Once random economic events select a particular path, the choice may become locked in regardless of the advantages of the alternatives."[13] And the creator of the QWERTY legend, Paul David, noted in reference to the consequences of the QWERTY design, "Competition in the absence of perfect futures markets drove the industry prematurely into standardization on the wrong system where decentralized decision making subsequently has sufficed to hold it."[14]

By this reasoning, if IBM gets to the personal computer market first, then other competitors are unfairly locked out. The economist Paul Krugman reasoned as much in his rendering of the QWERTY legend: "What computer do you use? Probably an IBM clone. Why do you use it? Probably because everyone else has an IBM clone. Does everyone else have an IBM clone because it's a superior product? Manifestly no. Why then? Because it got an early initial advantage, and there were increasing returns. What was the initial advantage? Thousands of corporate purchasing agents saying, for forty years, 'No one was ever fired for buying IBM.' Then there is the case of VHS vs. Beta. . . ."[15] By the same reasoning, if DOS is the first operating system, then competitors such as Macintosh's operating system will be unfairly locked out of the market. Apple cofounder Steve Wozniak made this very argument, in fact, citing the QWERTY legend: "Like the Dvorak keyboard, Apple's superior operating system lost the market-share war." Ergo, most economic inequalities are the result of unfair, chance-driven bottom-up markets that could be corrected with top-down measures. As Krugman concluded: "In QWERTY worlds, markets can't be trusted."

As we have just seen, however, we don't live in a QWERTY world. For example, all this assumes that IBM still holds market dominance in the PC business, which it doesn't, and that DOS was the first operating system, which it wasn't, and that users stuck grimly with DOS, which they didn't when they switched to Windows; and it is blind to the fact that virtually all operating systems today employ sleek graphical interfaces such as those originally provided by Mac. Not to mention the fact that computers replaced supposedly locked-in typewriters and that PCs displaced allegedly path-dependent mainframes.

This is not to say that history does not matter. Of course it does. Historical lock-in and path dependency are real phenomena. Note, for example, that we still use the "shift" key on our computer keyboards to produce a capital letter, even though there is no "carriage" inside the computer that needs shifting; and we still stroke the "return" key, even though there is no carriage to be returned to the start position. QWERTY may have a quirky history, but it has proved good enough to maintain its market dominance. If Dvorak (or some future alternative) proves superior to QWERTY, it will have to be superior enough to overcome personal and cultural momentum to achieve its own Nash equilibrium, Pareto optimum, or Evolutionary Stable Strategy. Technological systems, like biological ones, lock in their form and function according to a combination of efficacy and history. Optimal versus suboptimal is not the only deciding factor.

✦ ✦ ✦

Yet the "QWERTY reality" does not deep-six the concept of path dependency. We simply have to shift the emphasis of this biological model from the power of contingency to the potency of necessity. Balance and context are key.

The economic historian Douglas Puffert showed how to find that balance in his 2001 study on the standardization of railway track gauge.[16] The distance between a pair of rails seems fairly arbitrary, but establishing a standard that different railway companies or government agencies can abide by greases the wheels of commerce and produces lower costs, better service, and greater profits. Users of railway gauge are persuaded by the network effect of positive feedback loops to agree on a standard. Path dependency theory would predict that over time a single standard would be employed, but that is not the case. Australia and Argentina each have three different regional standard gauges, even though everyone recognizes that this is costing them all greater profits. India, Japan, Chile, and other countries each have two gauge standards that are used. Chalk one up for contingency. However, elsewhere, such as in the United States, Canada, and Great Britain, multiple gauges evolved into one commonly used standard. Chalk one up for necessity. (Though anyone who has traveled by train between Boston and Washington, D.C., will have discovered that while there is a single gauge standard for tracks, the most heavily trafficked passenger rail corridor still uses more than one engine system. Go figure.) In every case, regional markets determined the size and diversity of track gauge sizes as a function of both early design contingencies and later market efficiencies.

Contingencies are unmistakably influential in the early stages of a historical sequence, but they are quickly washed out by the oscillating waves of necessitating forces, which dig the dependent paths deeper and shore up their walls. As these necessitating forces gain in power, the paths reach points of equilibrium, optimization, and stability, where they hold sway for a time. But in the evolution of life as well as the history of technology, extinction is the rule, survival the exception. Path-dependent points of equilibrium, optimization, and stability can be disrupted, phase transitions from one state to another state realized, and life transitions and technological revolutions undertaken. Think of markets as oscillating through multiple peaks and valleys, where high points of equilibrium, optimization, and stability are reached, only to be knocked off their pedestals by competing forces of both contingency and necessity. All inventions have quirky beginnings. This does not guarantee them market immortality. What matters more than butterfly beginnings are the winds of consumer preference.

✦ ✦ ✦

One reason that so many of us have overstretched the parallels between evolution and economics is an all too apparent (once you look for it) difference between biological and cultural evolution. There are many real and fruitful parallels, as we have seen, but there are also important differences. The historian George Basalla identified them lucidly in his seminal work *The Evolution of Technology*.[17] Basalla begins by busting the myth of the inventor working in isolation, dreaming up new and innovative technologies out of sheer creative genius (the ping of the lightbulb flashing brilliantly in the mind). All technologies, Basalla claims, are developed out of either artifacts (artificial objects) or naturfacts (organic objects): "Any new thing that appears in the made world is based on some object already in existence." But some artifact had to be first—an invention that comes from no other invention, ex nihilo, as it were. If this is the case, then that artifact, Basalla shows, came from a naturfact.

As an example, one of the inventors of barbed wire, Michael Kelly, in 1868 claimed, "My invention [imparts] to fences of wire a character approximating to that of a thorn-hedge. I prefer to designate the fence so produced as a thorny fence."[18] Making an organic and evolutionary analogy, Basalla treats artifacts like organisms: "Artifacts are as important to technological evolution as plants and animals are to organic evolution."[19]

So where did we get the lone-genius myth? Basalla argues that it is a product of the state-enforced patent system that assumes one individual deserves all the credit for an invention, to the point where anyone else's claim to originality or further development can be prevented by force of law. The pace of technological progress is put in the hands of the government. Companies are induced to generate and purchase patents with no intention of further development—if they so choose—as a way of maintaining the technological status quo. Citing the Dutch system, which was patentless from 1869 to 1912, and the Swiss system, patentless from 1850 to 1907, Basalla shows that not only was neither Holland nor Switzerland hurt by this lack of top-down bureaucratic control over the creative process in technologies and markets, but they prospered more than ever before: "Switzerland experienced vigorous economic growth between 1850 and 1907. Industry was so successful that it attracted foreign capitalists who were willing to invest in new ventures despite the absence of patent protection."[20]

It might be argued that such laissez-faire systems are more appropriate for the nineteenth century than today, but Basalla cites a study that found seventy of the most important inventions of the first half of

the twentieth century were the result of independent inventors who did not patent their inventions: the automatic transmission, Bakelite, the ballpoint pen, cellophane, the cyclotron, the gyrocompass, insulin, the jet engine, Kodachrome film, magnetic recording, power steering, the safety razor, xerography, the Wankel rotary-piston engine, and the zipper.[21]

The tree of cultural history, of course, is not precisely analogous to the tree of evolutionary history. Whereas biological species are defined by their capacity to reproduce their own kind but are unable to inter-breed with others of a different kind, artifactual and ideological "species" can and do "interbreed" with others of radically different types. The artifactual tree allows branches to rejoin after separation, thereby spawning new species of technology. Biological change is Darwinian—inheritance of genes from one generation to the next. Cultural change is Lamarckian—the inheritance of acquired characteristics within a generation. The correspondence between technological and evolutionary systems is not perfectly one-to-one because biological systems are locked into their histories by relatively stable genetic programs, whereas technological systems are locked into their histories by relatively fluid cultural programs.

✦ ✦ ✦

One of Stephen Jay Gould's most important ideas about evolutionary change that does graft on well to technological change is his identification of an *exaptation*—a corollary to an adaptation—in which a feature that originally evolved for one purpose is later co-opted for a different purpose. Consider the not-so-humble wing. A fully formed, aerodynamically sound wing is a remarkable adaptation for flight that confers on its carrier such advantages as escaping from voracious predators, capturing elusive prey, and covering vast distances over rugged terrain or water. But what good is half a wing? For Darwinian gradualism to work, each successive stage of wing development would need to be functional. A little nub of a partial wing is not aerodynamically capable of flight, so wouldn't such mutations be selected against by nature? This argument, known as *the problem of incipient stages*, was made against Darwin, who

countered with a technology analogue: "Although an organ may not have been originally formed for some special purpose, if it now serves for this end we are justified in saying that it is specially contrived for it. On the same principle, if a man were to make a machine for some special purpose, but were to use old wheels, springs, and pulleys, only slightly altered, the whole machine, with all its parts, might be said to be specially contrived for that purpose. Thus throughout nature almost every part of each living being has probably served, in a slightly modified condition, for diverse purposes, and has acted in the living machinery of many ancient and distinct specific forms."[22]

Gould and his colleague Elizabeth Vrba called this solution an *exaptation*, and provided many examples of both naturfacts and artifacts.[23] In the case of wings, for example, the incipient stages had uses other than for aerodynamic flight. A half a wing was not a poorly developed wing but a well-developed something else. One candidate is thermoregulation, or the control of body temperature. In the fossil record, for example, the first feathers resemble hair and are similar to the insulating down of modern bird chicks.[24] Since it appears that modern birds descended from cold-blooded bipedal therapod dinosaurs, wings with feathers may have evolved to regulate body heat—holding them close to the body retains heat, stretching them out releases heat.[25] Yet another early adaptation for wings and feathers may have been to provide a boost to running, noted when it was discovered that some modern birds flap their wings to gain traction when running up steep inclines, even enabling them to climb straight up a 90-degree vertical structure.[26] Still another function for incipient wings on bipedal dinosaurs was grasping. The most famous transitional fossil in evolutionary history, *Archaeopteryx*, has wings whose surface area is large enough to support its body, asymmetrical feathers capable of attaining lift, and a shoulder that allows enough flexibility for an adequate upstroke of the wing necessary for flight. Nevertheless, *Archaeopteryx* retains many dinosaur features, including a functional grasping hand, for which the "wing" was probably originally adapted, and only later exapted for flight.[27] In biological evolution, structures can be adapted for one function and evolve into use for another function, or they may have multiple functions at one time.

Adaptations and exaptations can be found in artifacts as well as

naturfacts. Gould made the connection in an essay, "Tires to Sandals," in which he showed how people in third-world countries are often forced (by economic necessity) to exapt technologies originally adapted for other uses (there is even a template you can download online that shows exactly how to convert a tire into a pair of sandals).[28] In cycling technology, changes to pedals, helmets, gloves, and saddles also serve as examples of both adaptations and exaptations. The clipless pedal system, for example, was originally marketed by Look, Inc., a French company manufacturing ski bindings. The owner of Look, Bernard Tapie, also owned La Vie Claire, a professional cycling team with the two best cyclists in the world (Bernard Hinault and Greg LeMond). This economic arrangement produced a natural marriage of ski and cycling technology. The result was a radically new pedal system that did not look or perform in any way like the old toe-clip pedal system. Within a year virtually every professional rider was using these pedals, and within two years nearly all amateur cyclists had them on their bikes.

I witnessed firsthand the exaptation of modern helmet technology through my association with Bell Helmets, a race sponsor. The commonly used expanded-polystyrene helmets of today did not evolve from the old leather "hairnets" worn by cyclists for decades (and completely worthless in terms of absorbing the energy of an impact with the pavement). Rather, they were invented by Bell Helmet engineers who had been using the expanded-polystyrene technology for motorcycle helmets (where they were proved to work and approved by the appropriate government agencies), and expanded polystyrene itself was a technology developed for purposes having nothing to do with helmets. Yet the technology had to be modified for cyclists in order to attract a market. The old "Bell Shell" helmet, adequate for impact protection, had a reputation among racers as being worn only by "Freds" (the cycling equivalent of "geeks"). I told the engineers that the helmet had to look like a leather hairnet or else no serious cyclist would use it, and if serious cyclists wouldn't use it then amateurs wouldn't either. Thus, our first model—the *V1 Pro*—looked like a black leather hairnet on the outside but performed like a motorcycle helmet on the inside.

A final example of technological exaptation can be seen in gloves and saddles, the two primary points of contact (and discomfort) between rider and machine. For the previous century, leather was the material of choice for these technologies. The only change was variation in thickness, cut, design, and color. In the first Race Across America, in 1982, for example, all of us riders suffered badly from carpal tunnel syndrome in our hands, causing numbness and partial paralysis of the fingers, not to mention the painful saddle sores we developed from hundreds of hours of skin-on-leather contact. To assuage the pain and discomfort, about halfway through the race I began stuffing sponges into my gloves for padding, and I even tried an old European cycling trick of placing a raw steak in my cycling shorts to absorb road vibration. That year the race was televised on ABC's *Wide World of Sports*, and by chance a medical doctor in Waco, Texas, named Wayman Spence happened to watch with great interest because he had just developed a new gel technology used for bedridden patients who suffer from pressure sores. Dr. Spence promptly saw the parallels and brought me down to his small company in Waco—Spenco Medical—where Spence, a seamstress, and I took gloves and saddles and figured out how to put Spence's gel inside them to make cycling more comfortable. Within a few years nearly every glove and saddle company in the industry featured gel gloves, saddles, and saddle pads. They look like old glove and saddle technology, but they perform like new medical gel technology.[29]

◆ ◆ ◆

Like species, ideas and artifacts can develop not only by evolution (slow and gradual change), but also by revolution (sudden and dramatic change). We all build on the thoughts and ideas of those who came before, constructing and reconstructing as we try to push the frontiers forward. Some of these forward steps, however, are larger than others—small steps, giant leaps, and all that. Ideas and artifacts evolve, like living species, through descent with modification from their ancestors. Some make continuous steps while other make discontinuous leaps.

Continuities (evolution) indicate a contiguous and constant connection to the past, as change occurs gradually over time. A continuous mode of change is imperceptibly slow, as one thing gently blurs into another, as in the evolution of the modern racing bicycle from 1960 to 1980 (whose changes would be nearly invisible to the uninitiated). *Dis-*

continuities signify breaks from the past as change occurs suddenly and dramatically over time. A discontinuous mode of change is noticeably fast, as one thing jumps into another, such as the revolution of the racing bicycle from 1980 to 2000 (whose changes would be visible to anyone, aficionado or not). Thus, we may make the following definitions:

Continuity is a contiguous connection to the past that changes gradually over time.

Discontinuity is a discontiguous break from the past that changes suddenly and dramatically over time.

It is not that the racing bicycle did not change from 1960 to 1980. It did, but the changes were continuous—small improvements in efficiency on existing derailleur and braking systems, lighter metals for components, stronger frame construction methods and materials, lighter and more durable tires, different handlebar tape, and so on. The changes from 1980 to 2000, on the other hand, were noticeably more dramatic, and improvements often had as their source industries other than cycling. This change was discontinuous—completely new "index shifting" derailleur systems, totally different materials for frame construction (aluminum, titanium, graphite, carbon fiber, even plastic), aerodynamic designs, clipless pedals, expanded-polystyrene hard-shell helmets, gel gloves and saddles, and so on. Both "continuity" and "discontinuity" accurately describe the change in the bicycle, but as in the relative roles of contingency and necessity, it depends on the particular conditions present in the chronological sequence when the change is triggered. If dominated by powerful necessities, change will be slow and continuous. If dominated by unstable or competing necessities with influential contingencies, change will be rapid and discontinuous.

Granted that the basic bike design of the diamond frame, chain drive, and upright cyclist was continuous throughout the twentieth century, there was a convergence of inventions, events, and social conditions that led to the discontinuous leap starting in the early 1980s. I saw firsthand this shift from continuous to discontinuous change from the time I entered the sport in 1979 to the time I retired from racing a decade later. Bicycles in 1979 were better than bicycles in 1959, but not so much that an improvement in the rider could not easily erase the differences. Then a series of events occurred during the decade to trigger a discontinuous change:

- The 1979 gas crisis caused bicycle sales to jump significantly in a single year.
- The Japanese entered the American bicycle market, bringing their high-tech designs and philosophy of R&D-driven constant change for improvement, quickly surpassing the time-honored European companies such as Campagnolo.
- Cyclist Bryan Allan pedaled Paul MacCready's flying bicycle—the human-powered airplane *Gossamer Albatross*—across the English Channel, making world headlines.
- Speed skater Eric Heiden won five gold medals in the 1980 winter Olympics and credited bicycle racing for his success.
- The Southland Corporation, owners of the wildly successful chain of 7-Eleven convenience stores, hired Heiden and put together a world-class professional cycling team, pouring millions of dollars into the sport, including building two Olympic velodromes and promoting cycling as a sport for everyone.
- Former Los Angeles Dodger catcher John Marino hurt his back and took up cycling to recover, and loved the sport so much that he went on to break the transcontinental record by riding his bike across America in twelve days.
- An Academy Award–winning film centered on bicycle racing—*Breaking Away*—introduced America to the nuances of the sport (drafting, breaking away from the peloton, shaved legs, and especially imported Italian racing bikes and yellow Campagnolo cycling caps).
- The ecology movement pushed people into exploring cycling as a means of transportation free of fossil fuel consumption.
- A three-time Olympic bike racer named John Howard won the Hawaiian Ironman triathlon and was featured on ABC's *Wide World of Sports*.
- The International Human-Powered Vehicle Association held races with streamlined bicycles and bicycles fitted with aerodynamic devices (faired) that broke all previous (and long-held) speed rec-

ords because their rules allowed for any technological design (as long as it was human-powered).

- John Howard rode himself into the Guinness Book of World Records and onto Johnny Carson's *Tonight Show* by pedaling a bicycle behind a race car at 152 mph on the Bonneville Salt Flats.
- Francesco Moser, an Italian racer using special aerodynamic equipment, demolished the great Eddy Merckx's one-hour distance record, garnering considerable attention for more research on equipment, training, and, sorry to say, performance-enhancing drugs.
- The UCI's decision to allow Moser's record to stand (later revoked) triggered the United States Cycling Federation to employ Dr. Chester Kyle and other scientists to design the ultimate high-tech, lightweight racing bicycle for the 1984 summer Olympic games in Los Angeles. Ridden by scientifically trained cyclists (who had no Russian or East German competitors because of the boycott), a wealth of Olympic medals was collected by the U.S. team.
- Mountain bikes made riding more comfortable and accessible, as well as opening up another geographic region for cycling.
- An cyclist named Jonathan "Jock" Boyer became the first American to enter the prestigious Tour de France, finishing in twelfth place.
- Another American cyclist named Greg LeMond became the first to win the world championship road race and the Tour de France (three times) and as a consequence made the cover of *Sports Illustrated* as "Athlete of the Year."
- Cycling magazines, books, and videos promoted every aspect of the sport and machine, and industry trade shows exploded in size, every year introducing hundreds of new and improved technological gadgets that no cyclist could do without.

The convergence of these events, and others, combined to trigger a shift from continuous to discontinuous change.

✦ ✦ ✦

Exploring these differences in evolution versus technology tells us something deeper about cultural change. The trees of biological and cultural change diverge even more deeply in terms of what drives each system.

Biological species are driven primarily by a need to survive, which includes food, water, reproduction, shelter from the elements, protection from predators, and so forth. Humans, as biological species, have all these needs, but we have more, many more. If necessity were the only mother of invention, then 99.9 percent of all inventions would never have come to fruition. What need, really, is there for typewriters, bicycles, airplanes, and computers? What did people do before there were such inventions?

Our basic needs have always been there. What we developed were new *wants*. Here is how it works: in evolution, genotypes generate phenotypes of a wide diversity upon which natural selection acts; that is, genetic variation produces a vast array of physical varieties, and nature selects those that are most fit to survive, as defined by passing on those genes into the next generation through different reproductive success, or leaving behind offspring. In economics, artifactual diversity and novelty provide the substance of this technological natural selection, in which consumers are selecting those products they deem most fit to survive, as defined by continual manufacturing, sales, and use. This artifactual system is driven by the deeply ingrained human desire for a richer and more fulfilling existence. The artifacts in our made world do not just represent narrow solutions to problems that arose out of our basic physical needs, but constitute material expressions of our psychological aspirations and spiritual needs. This desire for a more fulfilling existence is what it means to be human. We build typewriters and computers, bicycles, and automobiles, not so much because we need them but because we want them.

5

MINDING OUR MONEY

In December 1954, the psychologist Leon Festinger and his colleagues noticed this newspaper headline: PROPHECY FROM PLANET CLARION CALL TO CITY: FLEE THAT FLOOD. A Chicago housewife, Marion Keech, reported that she had received messages from the planet Clarion telling her that the world would end in a great flood sometime before dawn on December 21, 1954. If she and her followers gathered together at midnight, however, a mother ship would arrive just in time to whisk them away to safety.

Festinger immediately saw an opportunity, not to save himself, but to study the phenomenon of *cognitive dissonance*, the mental tension created when a person holds two conflicting thoughts simultaneously. "Suppose an individual believes something with his whole heart," Festinger said. "Suppose further that he has a commitment to this belief, that he has taken irrevocable actions because of it; finally, suppose that he is presented with evidence, unequivocal and undeniable evidence, that his belief is wrong: what will happen? The individual will frequently emerge, not only unshaken, but even more convinced of the truth of his beliefs than ever before. Indeed, he may even show a new fervor about convincing and converting other people to his view."[1] Many of Keech's followers had quit their jobs, left their spouses, and given away their possessions. Festinger predicted that these individuals with the strongest behavioral commitment would be the *least* likely to admit their error when the prophecy failed and instead rationalize a positive outcome.

As midnight approached on December 20, Keech's group gathered to await the arrival of the aliens' mother craft. As dictated by Marion, the members eschewed all metallic items and other objects that would interfere with the operation of the spaceship. When one clock read 12:05 A.M. on the twenty-first, anxious squirming was calmed when someone pointed out a second clock reading 11:55 P.M. But as the minutes and hours ticked by, Keech's clique grew restless.

At 4:00 A.M., Keech began to weep in despair, recovering at 4:45 A.M. with the claim that she had received another message from Clarion informing her that God had decided to spare Earth because of the cohort's stalwart efforts. "By dawn on the 21st, however, this semblance of organization had vanished as the members of the group sought frantically to convince the world of their beliefs," Festinger says. "In succeeding days, they also made a series of desperate attempts to erase their rankling dissonance by making prediction after prediction in the hope that one would come true, and they conducted a vain search for guidance from the Guardians."[2] Marion Keech and her most devoted charges redoubled their recruitment efforts, arguing that the prophecy had actually been fulfilled with an opposite outcome as a result of their faith. Festinger concluded that Keech's assemblage reduced the cognitive dissonance they experienced by reconfiguring their perceptions to imagine a favorable outcome, reinforced by converting others to the cause.[3]

Doomsday cults are especially vulnerable to cognitive dissonance, particularly when they make specific end-of-the-world predictions that will be checked against reality. What typically happens is that the faithful spin-doctor the nonevent into a successful prophecy, with rationalizations including (1) the date was miscalculated; (2) the date was a loose prediction, not a specific prophecy; (3) the date was a warning, not a prophecy; (4) God changed his mind; (5) the prediction was just a test of the members' faith; (6) the prophecy was fulfilled physically, but not as expected; and (7) the prophecy was fulfilled—spiritually.[4]

Of course, cognitive dissonance is not unique to doomsday cults. We experience it when we hang on to losing stocks, unprofitable investments, failing businesses, and unsuccessful relationships. Why should past investment influence future decisions? If we were perfectly rational, we should simply compute the odds of succeeding from this point forward, jettisoning our previous beliefs. Instead, we are stuck rationalizing our past choices, and those rationalizations influence our present ones.

+   +   +

Unfortunately for those bent on curing themselves of the chronic effects of cognitive dissonance, research since Festinger shows that, if anything, he underestimated its potency. As two of Festinger's students—Carol Tavris and Elliot Aronson—demonstrate in their aptly titled book, *Mistakes Were Made (but not by me)*, our ability to rationalize our choices and actions through *self-justification* knows no bounds.

The passive voice of the all-telling phrase—*mistakes were made*—shows the rationalization process at work. In March 2007, United States attorney general Alberto R. Gonzales used that very phrase in a public statement on the controversial firing of several U.S. attorneys: "I acknowledge that mistakes were made here. I accept that responsibility." Nevertheless, he rationalized, "I stand by the decision, and I think it was a right decision."[5] The phraseology is so common as to be almost cliché. "Mistakes were quite possibly made by the administrations in which I served," confessed Henry Kissinger about Vietnam, Cambodia, and South America. "If, in hindsight, we also discover that mistakes may have been made . . . I am deeply sorry," admitted Cardinal Edward Egan of New York about the Catholic Church's failure to deal with priestly pedophiles. And, of course, corporate leaders are no less susceptible than politicians and religious leaders: "Mistakes were made in communicating to the public and customers about the ingredients in our French fries and hash browns," acknowledged a McDonald's spokesperson to a group of Hindus and other vegetarians after they discovered that the "natural flavoring" in their potatoes contained beef byproducts. "Dissonance produces mental discomfort, ranging from minor pangs to deep anguish; people don't rest easy until they find a way to reduce it," Tavris and Aronson note.[6] It is in that process of reducing dissonance that our self-justification accelerators are throttled up.

One of the practical benefits of self-justification is that no matter what decision we make—to take this or that job, to marry this or that person, to purchase this or that product—we will almost always be satisfied with the decision, even when the objective evidence is to the contrary. Once the decision is made, we carefully screen subsequent information and filter out all contradictory data, leaving only evidence in support of our choice. This process of cherry-picking happens at even the highest levels of expert assessment. In his book *Expert Political Judgment*, the political scientist Philip Tetlock reviews the evidence for the predictive ability of professional experts in politics and economics and finds them severely wanting. To the point, expert opinions turn out to be no better than those of nonexperts—or even chance—and yet, as self-justification theory would predict, experts are significantly less likely to admit that they are wrong than are nonexperts.[7]

Politics is rampantly self-justifying. Democrats see the world through liberal-tinted glasses, while Republicans filter it through conservative-shaded lenses. Tune in to talk radio any hour of the day, any day of the week—whether it is "conservative talk radio" or "progressive talk radio"— and you'll hear the same current events interpreted in ways that are 180 degrees out of phase. Social psychologist Geoffrey Cohen quantified this effect in a study in which he discovered that Democrats are more accepting of a welfare program if they believe it was proposed by a fellow Democrat, even when, in fact, the proposal comes from a Republican and is quite restrictive. Predictably, Cohen found the same effect for Republicans, who were far more likely to approve a generous welfare program if they thought it was proposed by a fellow Republican.[8]

Economic positions, whether staked out by birthright, inheritance, or creative hard work, distort our perceptions of reality as much as political positions. The sociologist John Jost has studied how people justify their economic status, and the status of others. The wealthy tend to rationalize their position of privilege as deserved, earned, or justified by their benevolent social acts, and assuage any cognitive dissonance regarding the poor by believing that the poor are happier and more honest. For their part, the underprivileged tend to rationalize their position as morally superior, nonelitist, and within the bounds of social normalcy, and look down upon the rich as living an undeserved life of accidental or ill-gotten privilege.[9]

Cognitive distortions can even turn deadly. Wrongly convicting people and sentencing them to death is a supreme source of cognitive dissonance. Since 1992, the Innocence Project has freed fourteen people from death row, and exonerated convicts in more than 250 non-death-row cases. "If we reviewed prison sentences with the same level of care that we devote to death sentences," says University of Michigan law professor Samuel R. Gross, "there would have been *over 28,500 non-death-row exonerations in the past 15 years*, rather than the 255 that have in fact occurred." What is the self-justification for reducing this form of dissonance? "You get in the system and you become very cynical," explains Rob Warden of Northwestern University School of Law. "People are lying to you all over the place. Then you develop a theory of the crime, and it leads to what we call tunnel vision. Years later, overwhelming evidence comes out that the guy was innocent. And you're sitting there thinking, 'Wait a minute. Either this overwhelming evidence is wrong or I was wrong—and I couldn't have been wrong because I'm a good guy.' That's a psychological phenomenon I have seen over and over."[10]

The deeper evolutionary foundation to self-justification, cognitive dissonance, and the elevation of truth telling and mistake admission to a moral principle worthy of praise can be found in the psychology of deception (and self-deception). Research shows that we are better at deception than at deception detection, but liars get caught often enough that it is risky to attempt to deceive others, especially people with whom we spend a lot of time. The more we interact with someone, the more that person is likely to pick up on the cues we give when we are attempting to deceive, particularly nonverbal cues such as taking a deep breath, looking away from the person we are talking to, and hesitating before answering. But those cues are less likely to be expressed if you actually believe the lie yourself.[11] This is the power of self-deception, which evolved in our ancestors as a means of fooling fellow group members who would otherwise catch our deceptions.

From an evolutionary perspective, it is not enough to fake doing the right thing, because although we are fairly good deceivers, we are also fairly good deception detectors. We have to believe we are doing the right thing, too. What we believe we feel, and thus it is that we do not just go through the motions of being moral, we actually have a moral sense and retain the capacity for genuine moral emotions. This is borne out in research on both primates and hunter-gatherer groups, as we shall see in the next chapter. What follows are the numerous ways that cognitive biases interrupt our ability to make rational decisions in our personal as well as our financial lives.

✦ ✦ ✦

Picture yourself watching a one-minute video of two teams of three players each, one team wearing white shirts and the other black shirts, as they move about each other in a small room tossing two basketballs back and forth among themselves. Your task is to count the number of passes made by the white team. Unexpectedly, after thirty-five seconds a gorilla enters the room, walks directly through the farrago of bodies, thumps his chest, and nine seconds later, exits. Would you see the gorilla?

Most of us, in our perceptual vainglory, believe we would—how could anyone miss a guy in an ape suit? In fact, 50 percent of subjects in this remarkable experiment by psychologists Daniel Simons and Christopher Chabris do not see the gorilla, even when asked if they noticed anything unusual.[12] The effect is known as *inattentional blindness*—when attending to one task, say, talking on a cell phone while driving, many of us become blind to dynamic events, such as a gorilla in the crosswalk. For several years now, I have incorporated the gorilla DVD into my public lecture on "The Power of Belief," asking at the end of the talk for a show of hands of those who did not see the gorilla. Out of the more than one hundred thousand people over the years, fewer than half saw the gorilla. I can decrease the figure even more by issuing a gender challenge, telling the audience before showing the clip that one gender is more accurate than the other at counting the ball passes, but I won't tell them which gender so as not to bias the test. This really makes people sit up and concentrate, causing even more to miss the gorilla. The lowest percentage I have ever witnessed miss the gorilla was group of about fifteen hundred behavioral psychologists. Professional observers of behavior, almost none of them saw the gorilla. Many were shocked. Several accused me of showing two different clips.[13]

Experiments such as these reveal a hubris in our powers of perception, as well as a fundamental misunderstanding of how the brain works.

We think of our eyes as video cameras, and our brains as blank tapes to be filled with precepts. Memory, in this flawed model, is simply rewinding the tape and playing it back in the theater of the mind, in which some cortical commander watches the show and reports to a higher homunculus what it saw. Fortunately for criminal defense attorneys, this is not the case. The perceptual system, and the brain that analyzes its data, are far more complex. As a consequence, much of what passes before our eyes may be invisible to a brain focused on something else.

Driving is an example. "Many accident reports include claims like 'I looked right there and never saw them,'" Simons told me. "Motorcyclists and bicyclists are often the victims in such cases. One explanation is that car drivers expect other cars but not bikes, so even if they look right at the bike, they sometimes might not see it." Simons recounted for me a study by Richard Haines of pilots who were attempting to land a plane in a simulator with the critical flight information superimposed on the windshield. "Under these conditions, some pilots failed to notice that a plane on the ground was blocking their path."[14] There are none so blind as those who will not see.

✦   ✦   ✦

Have you ever noticed how blind other people are to their own biases, but how you almost always seem to catch yourself before falling for your own? If so, then you are the victim of the *blind spot bias*, in which subjects recognized the existence and influence in others of eight different cognitive biases but failed to see those same biases in themselves. In one study, Stanford University students were asked to compare themselves to their peers on such personal qualities as friendliness and selfishness. Predictably, they rated themselves higher than they rated their peers. When the subjects were warned about the *better-than-average bias* and asked to reevaluate their original assessments, 63 percent claimed that their initial evaluations were objective, and 13 percent even claimed to have been too modest! In a related study, Princeton University psychologist Emily Pronin and her colleagues randomly assigned subjects high or low scores on a "social intelligence" test.

Unsurprisingly, those given the high marks rated the test fairer and more useful than did those receiving low marks. When asked if it was possible that they had been influenced by the score they received on the test, subjects responded that other participants had been far more biased than they were. When subjects admit to having such a bias as being a member of a partisan group, says Pronin, this "is apt to be accompanied by the insistence that, in their own case, this status . . . has been uniquely *enlightening*—indeed, that it is the *lack* of such enlightenment that is making those on the other side of the issue take their misguided position."[15]

In a third study in which Pronin queried subjects about what method they used to assess their own and others' biases, she found that people tend to use general theories of behavior when evaluating others, but use introspection when appraising themselves. The problem with this method can be found in what Pronin calls the *introspection illusion*, in which people trust themselves to employ the subjective process of introspection but do not believe that others can be trusted to do the same.[16] Okay for me but not for thee. "We view our perceptions of our mental contents and processes as the gold standard for understanding our actions, motives, and preferences," Pronin explained to me. "But, we do not view others' perceptions of their mental contents and processes as the gold standard for understanding their actions, motives, and preferences. This 'illusion' that our introspections are a gold standard leads us to introspect to find evidence of bias and we are thus likely to infer that we have not been biased, since most biases operate outside of conscious awareness."[17]

We tend to see ourselves in a more positive light than others see us. National surveys show that most businesspeople believe they are more moral than other businesspeople,[18] while psychologists who study moral intuition think they are more moral than other such psychologists.[19] In one College Entrance Examination Board survey of 829,000 high school seniors, 60 percent put themselves in the top 10 percent in "ability to get along with others," while 0 percent (not one!) rated themselves below average.[20]

This *self-serving bias* is Lake Wobegon writ large. An amusing example of it can be found in a 1997 *U.S. News & World Report* study on who Americans believe is likely to go to heaven: 52 percent said Bill Clinton, 60 percent thought Princess Diana, 66 percent chose Oprah Winfrey, and 79 percent selected Mother Teresa. But 87 percent chose themselves as the person most likely to go to heaven![21]

✦ ✦ ✦

What people see in others they generally do not see in themselves. This leads to an *attribution bias*. A number of studies show that there is a tendency for people to accept credit for their good behaviors (a dispositional judgment) and to allow the situation to account for their bad behaviors.[22] In dealing with others, we are more likely to attribute both good and bad actions to dispositional factors. Hence, we tend to attribute our own good fortune to hard work and intelligence, whereas another person's good fortune is the result of luck and circumstance. Conversely, our own bad actions are a product of circumstance, but we blame the bad actions of others on their personal weaknesses.[23]

My colleague Frank J. Sulloway, a psychologist and historian of science at the University of California, Berkeley, and I have discovered another bias in how we assess our behavior and that of others. We wanted to know why people believe in God, so we polled ten thousand randomly selected Americans. In addition to exploring various demographic and sociological variables, we directly asked subjects to respond in writing to two open-ended questions: Why do you believe in God? and Why do you think others believe in God? The top two reasons that people gave for why they believe in God were "the good design of the universe" and "the experience of God in everyday life." Interestingly, and tellingly, when asked why they think other people believe in God, these two answers dropped to sixth and third place, respectively, while the two most common reasons given were that belief is "comforting" and "fear of death."[24] There appears to be a sharp distinction between how people view their own beliefs—as rationally motivated—and how people view the beliefs of others—as emotionally driven. A person's commitment to a belief is generally attributed to an intellectual choice ("I bought these $200 jeans because they are exceptionally well made and fit me perfectly" or "I am for gun control because statistics show that crime decreases when gun ownership decreases"), whereas another person's decision or opinion is attributed to need or emotional reasons ("She bought those overpriced designer jeans because she is obsessed with appearing hip and matching the status-hungry in-crowd," or

"He is for gun control because he is a bleeding-heart liberal who needs to identify with the victim").[25]

✦ ✦ ✦

Given the ubiquity and power of such cognitive biases, it was only a matter of time before someone created an entire branch of economics to study them. The field, *behavioral economics*, was pioneered by a couple of psychologists, Daniel Kahneman and Amos Tversky, neither of whom ever took a single course in economics but whose personal experiences in war and scientific training in how the mind works have led to one of the most fruitful collaborations in the history of social science.

Daniel Kahneman was born in Tel Aviv, grew up in Paris, and returned to his homeland shortly after the Second World War. He later recalled that his interest in the complexities and inconsistencies of human behavior was shaped by a salient encounter with a Nazi SS officer in France shortly after the Nazi occupation. Forced to wear the yellow Star of David, the young boy was returning home past curfew one evening, his sweater turned inside out, when a German soldier approached. Kahneman tried to scurry past, but "he beckoned me over, picked me up, and hugged me. I was terrified that he would notice the star inside my sweater. He was speaking to me with great emotion, in German. When he put me down, he opened his wallet, showed me a picture of a boy, and gave me some money. I went home more certain than ever that my mother was right: people were endlessly complicated and interesting."[26] Kahneman earned his doctorate in psychology from the University of California, Berkeley, going on to win the Nobel Prize in economics in 2002.

Amos Tversky would have won the prize as well, but in 1996 he died of metastatic melanoma at the age of fifty-nine. Also Israeli-born and keenly interested in understanding the subtleties and oddities of human behavior, Tversky later recalled that "growing up in a country that's fighting for survival, you're perhaps more likely to think simultaneously about applied and theoretical problems." "Applications" included a stint as a paratrooper in an elite unit in the Israeli army, earning Tversky his country's highest honor for bravery during a 1956 border skirmish in which a fellow soldier had fallen on top of an armed explosive device he was placing beneath some barbed wire. Following a few steps behind the soldier, without thinking and against the orders of his commanding officer to stay put, Tversky leaped forward and pulled the man off the explosive device, thereby saving him. Tversky was wounded and lived the rest of his life with shards of metal in his body, but the lesson was imparted: people do not always make rational choices.

Tversky noticed that people are pattern-seeking animals, finding meaningful relationships in random data, everything from stock market fluctuations to coin tosses to sports streaks. A basketball fan, Tversky teamed with fellow cognitive psychologists Robert Vallone and Thomas Gilovich to test the notion of "hot hands." As fans of the game know, when you're hot you're hot, and when you're not you're not, and you can see it every night on the court. Knowing the propensity for humans to find such patterns whether they are there or not, Tversky and his colleagues analyzed every shot taken by the Philadelphia 76ers basketball team for an entire season, only to discover that the probability of a player hitting a second shot after a successful basket did not increase beyond what one would expect by chance and by the average shooting percentage of the player. That is, the number of successful baskets in sequence did not exceed the predictions of a statistical coin-flip model. If you conduct a coin-flipping experiment and record heads or tails, you will encounter streaks: on average and in the long run, you will flip five heads or tails in a row once in every thirty-two sequences of five tosses. Most of us will interpret these random streaks as meaningful.[27] Indeed, it would be counterintuitive not to do so.

It was Tversky's research in cognitive psychology that led him to challenge the dogma in economic theory that people act rationally to maximize their welfare. According to the economist Kenneth Arrow, one of the founders of neoclassical economics and the youngest person ever to make the trip to Sweden to collect The Prize, "Previous criticism of economic postulates by psychologists had always been brushed off by economists, who argued, with some justice, that the psychologists did not understand the hypotheses they criticized. No such defense was possible against Amos' work."[28] By most accounts, Tversky was a genius.

Yet, with characteristic modesty, Tversky said that most of his findings in economics were well known to "advertisers and used car salesmen." For example, he noticed that customers were displeased when a retail store added a surcharge to a purchase made with a credit card, but were delighted when the store instead offered a discount for paying in cash. This "framing" of a choice as a penalty or as a reward deeply influences one's decision. In another example, Tversky discovered that in reviewing the risks of a medical procedure with patients, physicians will get a very different response if they tell them that there is a 1 percent chance of dying versus a 99 percent chance of surviving.[29]

Kahneman and Tversky, in conjunction with their colleagues Richard Thaler, Paul Slovic, Thomas Gilovich, Colin Camerer, and others, established a research program to study the cognitive basis for common human errors in thinking and decision making. They discovered a number of "judgmental heuristics," as they called them—"mental shortcuts," or simpler still, "rules of thumb"—that shape our thinking, most especially how we think about money.

Imagine that you work for the admissions office of a graduate program at a university and you come across this description of a candidate in a letter of recommendation:

> Tom W. is of high intelligence, although lacking in true creativity. He has a need for order and clarity, and for neat and tidy systems in which every detail finds its appropriate place. His writing is rather dull and mechanical, occasionally enlivened by somewhat corny puns and by flashes of imagination of the sci-fi type. He has a strong drive for competence. He seems to feel little sympathy for other people and does not enjoy interacting with others. Self-centered, he nonetheless has a deep moral sense.

Kahneman and Tversky presented this scenario to three groups of subjects. One group was asked how similar Tom W. was to a student in one of nine types of college graduate majors: business administration, computer science, engineering, humanities/education, law, library science, medicine, physical/life sciences, or social science/social work. Most of the subjects in the group associated Tom W. with an engineering student, and thought he was least like a student of social science/social work. A second group was asked instead to estimate the probability that Tom W. was a grad student in each of the nine majors. The probabilities lined up with the judgments from the first group. A third group was asked to estimate the proportion of first-year grad students in each of the nine majors. What the researchers found was that even though the subjects—based on the answers of the third group, or control—knew that there are far more graduate students in the social sciences than there are in engineering, and thus the probability of Tom W.'s being in the engineering program is the lowest, they nevertheless concluded that he must be an engineer based on what the narrative description of him represented.[30]

Tversky and Kahneman call this the *representative fallacy*, in which "an event is judged probable to the extent that it represents the essential features of its parent population or generating process." And, more generally, "when faced with the difficult task of judging probability or frequency, people employ a limited number of heuristics which reduce these judgments to simpler ones."[31] We are good at telling stories about people and gleaning from these narratives bits of information that we then use to make snap decisions.

On the other hand, we are lousy at computing probabilities and thinking in terms of the chances of something happening. This time, imagine that you are looking to hire someone for your company and you are considering the following candidate:

> *Linda is 31 years old, single, outspoken, and very bright. She*
> *majored in philosophy. As a student, she was deeply concerned with*
> *issues of discrimination and social justice, and also participated in*
> *antinuclear demonstrations.*

Which is more likely? (1) Linda is a bank teller, or (2) Linda is a bank teller and is active in the feminist movement.

When this scenario was presented to subjects, 85 percent chose the second option. Mathematically speaking, this is the wrong choice, simply because the probability of two events occurring together (in "conjunction") will always be less than or equal to the probability of either one occurring alone. Tversky and Kahneman argue that most people get this problem wrong because the second option appears to be more "representative" of the description of Linda.[32]

Hundreds of experiments in these fallacies have been conducted, showing time and again that people make snap decisions under high levels of uncertainty, and they do so by employing these various rules of thumb to shortcut the computational process. For example, policy experts were asked to estimate the probability that the Soviet Union would invade Poland and that the United States would then break off diplomatic relations. Subjects gave the scenario a probability of 4 percent. Meanwhile, another group of policy experts was asked to estimate the probability just that the United States would break off diplomatic relations with the Soviet Union. Contrary to what the odds actually would be, these experts gave this latter scenario only a 1 percent chance of happening. The experimenters concluded that the more detailed two-part scenario seemed more likely because we like stories with more details and thus grant them greater veracity.

◆ ◆ ◆

In the run-up to the 1976 U.S. presidential election, an experiment was conducted in which one group of subjects was asked to "imagine Gerald Ford winning the upcoming election," while another group of subjects was asked to "imagine Jimmy Carter winning the upcoming election." When subsequently asked to estimate the probability of each of the candidates winning, those who were asked to imagine Ford winning estimated his chances as much higher than those who were asked to imagine Carter winning, who, in turn, gave their guy a much higher probability of victory.[33]

This is the *availability fallacy*, which holds that we assign probabilities of potential outcomes based on examples that are immediately available to us, which are then generalized into conclusions upon which choices are based.[34] Your estimation of the probability of hitting red lights during a drive will be directly related to whether you are late or not. Your estimation of the probability of dying in a plane crash (or lightning strike, shark attack, terrorist attack, etc.) will be directly related to the availability of just such an event in your world, especially your exposure to it in the media. If newspapers cover an event, there is a good chance that people will overestimate the probability of that event happening.[35]

The USC sociologist Barry Glassner has built his career studying how the media creates a "culture of fear" that is totally out of sync with reality. Because the media establishes an availability rule of thumb about what we should fear, we fear all the wrong things and ignore those that could potentially kill us. "In the late 1990s the number of drug users had decreased by half compared to a decade earlier," Glassner notes, yet the "majority of adults rank drug abuse as the greatest danger to America's youth."

The availability distortion of our understanding of medical issues is especially egregious. Studies have found that women in their forties believe they have a 1 in 10 chance of dying from breast cancer, while their real lifetime odds are more like 1 in 250. This effect is directly related to the number of news stories about breast cancer. As Glassner documents, "We waste tens of billions of dollars and person-hours every year on largely mythical hazards like road rage, on prison cells occupied by people who pose little or no danger to others, on programs designed to protect young people from dangers that few of them ever face, on compensation for victims of metaphorical illnesses, and on technology to make airline travel—which is already safer than other means of transportation—safer still."[36] Our perceptions of the economy are similarly distorted: "the unemployment rate was below 5 percent for the first time in a quarter century. Yet pundits warned of imminent economic disaster."

Available information can distort our decision making in surprising ways. Two groups of subjects are asked to supply an estimate:

Group 1: What percentage of African nations are members of the United Nations? Do you think it is more or less than 45 percent? Please give an exact percentage.

Group 2: What percentage of African nations are members of the United Nations? Do you think it is more or less than 65 percent? Please give an exact percentage.

Subjects who answer the first question give a lower percentage estimate than subjects who answer the second question. Why? They were given a lower starting point, which primed their brains to think in lower numbers. Once an initial value is set, we are biased toward that value. Behavioral economists call this effect the *anchoring* fallacy, and it shapes our perceptions of what we consider to be a fair price or a good deal. After all, money is simply cheap paper with some ink slapped on it, so the value of a commodity must be evaluated in a context. The context begins with an anchor. Lacking some objective standard, which is usually not available, we grasp for any standard available, no matter how seemingly subjective. It reminds me of that old Henny Youngman routine: "How's your wife?" "Compared to what?"

The comparison anchor can even be entirely arbitrary. In one study, subjects were asked to give the last four digits of their Social Security numbers, and then asked to estimate the number of physicians in New York City. Bizarrely, people with higher Social Security numbers tended to give higher estimates for the number of docs in Manhattan. In a related experiment, subjects were shown an array of items to purchase—a bottle of wine, a cordless keyboard computer, a video game—and were then told that the price of the items was equal to the last two digits of their Social Security numbers. When subsequently asked the maximum price they would be willing to pay, subjects with high Social Security numbers consistently said that they would be willing to pay more than those with low numbers.

Our intuitive sense of the anchoring effect and its power explains why negotiators in corporate mergers, representatives in business deals, and even those involved in divorces stand to benefit from beginning from an extreme initial position. Setting a high anchor mark influences the information "available" to both sides. But once an event has occurred, it is easy to look back and reconstruct not only how it happened—not only why it had to happen that way and not some other way—but also why we should have seen it coming all along. Known colloquially as "Monday morning quarterbacking," the *hindsight bias* is the tendency to reconstruct the past to fit present knowledge.[37] The hindsight bias went to work after December 7, 1941, when it became clear after the fact that the Japanese had always planned to attack Pearl Harbor. The proof was in the so-called "bomb plot message," intercepted in October 1941 by U.S. intelligence, in which a Japanese agent in Hawaii was instructed by his superiors in Japan to monitor warship movements in and around the Oahu naval base. The "fateful" message was never passed up to President Roosevelt. There were, in fact, eight such messages dealing with

Hawaii as a possible target that were intercepted and decrypted by intelligence agents before the attack. In hindsight and out of context, it looked like a terrible failure. In context and without hindsight, however, it was not at all clear where the Japanese were going to strike. In fact, from May through November 1941, army intelligence, concerned about security leaks and the possibility that the Japanese might discover that their codes had been broken, stopped sending all such memos to the White House. More critical, during the same period in which the eight messages involving ship movements in and around Hawaii were intercepted, no fewer than fifty-eight ship movement messages were intercepted in association with the Philippines, twenty-one involving Panama, seven regarding Southeast Asia and the Netherlands East Indies, and even seven connected to the U.S. West Coast![38]

In like manner, one of the only predictable stock market effects is the emergence of clear hindsight the day after the market does anything out of the ordinary—moving up or down—regardless of the actual causes. In conditions of causal uncertainty, our cognitive biases kick into gear and drive us to concoct all sorts of probable causes. During the time I was writing this book, Google stock, as it is wont to do, took several major plunges, only to bounce right back. When the stock rose dramatically after a positive quarterly earnings report, the Monday morning money pundits proclaimed that investors were rewarding Google for a job well done. Yet the next quarterly earnings report, which was even more spectacular than the previous one that garnered investors' favor, resulted in the stock's plunging more than $40 in a matter of days. The cause was obvious to the stock analysts: investors were punishing Google for not increasing their quarterly earnings as much as these same analysts predicted that they would.

We connect the dots from our complex and seemingly chaotic world and construct narratives based on the connections we think we have found. Whether the patterns are real or not is a separate issue entirely. In neoclassical economics, however, the assumption was that no matter how muddy our storytelling became, the data would never lie. What economists had tried to ignore before Tversky and Kahneman was that we decide what the data say in the first place—and our brains evolved to deal with a world that bears only slight resemblance to the vast, messy crowds of information in the modern marketplace.

✦ ✦ ✦

Before you sit two black bags filled with red and white marbles. In one of the bags, two-thirds of the marbles are red and one-third are white, while in the other bag one-third of the marbles are red and two-thirds are white. Your mission is to figure out which bag has the most red marbles and which bag has the most white marbles. You are allowed to pull five marbles out of bag number one and thirty marbles out of bag number two. From the first bag, you grab four red marbles and one white marble. From the second bag, you grab twenty red marbles and ten white marbles. Which bag would you predict has the most red marbles? Most people would say the first bag, because four out of five red marbles represents 80 percent, whereas twenty out of thirty red marbles in the second bag represents only 66 percent. But statistically speaking, the smarter bet is bag two, because the sample size is larger and therefore more representative of the total number of red marbles.[39]

Because of the *law of small numbers* we tend to believe that small sample sizes are representative of the larger population. Investors make mistakes of this sort on a daily basis, when a short-term rise or fall in a stock is taken as deeply meaningful and thus serves as a trigger to buy or sell. The wise but counterintuitive approach would be to first call up those long-term trend charts to see if the sawtooth daily ups and downs of the stock are part of a trend up or down. Like global temperatures, they rise and fall daily, so basing decisions on small numbers of data points is unwise.

A corollary to the law of small numbers is that if your numbers are large enough, then a random and representative sample from it will give you a good sense of the way the world really is. In science, if you run an experiment over and over, the observed probability will approach the real (or actual) probability. This is one meaning of the *law of large numbers*, but another one that I find useful in explaining odd occurrences is that if the numbers are large enough, the probability is that something weird is likely to happen. That is, very improbable events will probably occur when there are a sufficiently large number of chances for it to happen.

Million-to-one odds happen three hundred times a day in America. With three hundred million Americans running around doing their thing each day, it is inevitable that on any given nightly newscast, there will be a story about something really weird happening to someone somewhere.

When it comes to the modern complex economy, which involves the manufacturing, distribution, and sale of billions of products traded by billions of people, weird things will happen on a fairly regular basis. For instance, compare the "most traded" stocks on any given day to the "biggest gainers" and "biggest losers" on those same days: they are never the same companies. The most traded stocks represent the largest and most popular companies with the most shares available to buy and sell, whereas the biggest gainers and losers almost always consist of companies you will never have heard of because they are the outliers, the extremes. Because there are now so many companies whose stock is publicly traded, the law of large numbers practically guarantees that on any given day someone is going make a huge amount of money while someone else is going to lose a large sum of cash. How can the human mind successfully manage choices with probabilities in this context?

As it happens, it isn't very good at probabilities in the context of small numbers, either. On the classic television game show *Let's Make a Deal*, contestants were forced to choose one of just three doors. Behind one of the doors the contestant would find (and win!) a brand-new car. Behind the other two doors the contestant would discover goats (and not even brand-new ones). If a contestant chose door number one and host Monty Hall, who knew what was behind all three doors, then showed her a goat behind door number two, would the contestant be smarter to stick with door number one or switch to door number three?

Most people figure that it doesn't matter because now the odds are fifty-fifty. But in this situation, most people would be wrong. In fact, in this particular real-life example (first presented by Marilyn vos Savant in her weekly *Parade* magazine column), "most people" included not just the general public but scientists, mathematicians, and even some statisticians, who upbraided her in no uncertain terms for the ignorance of her ways.

Yet all of their intuitions and rationalizations, expert and otherwise, were misguided. Here's the explanation: the contestant had a one in three chance before any door was opened, but once Monty shows one of the losing doors, she has a two-thirds chance of winning by switching. Why? There are three scenarios for the doors: (1) car goat goat; (2) goat car goat; (3) goat goat car. If the contestant faces scenario one, in which the car is behind the door she has chosen already, she loses by switching, but if she faces either scenario two or three, she will win by switching. Another way to reason around this counterintuitive problem in probabilities is to imagine that there are a hundred doors, not just three; the contestant chooses door number one and Monty shows her doors number two through ninety-nine—all goats. Now should she switch? Of course she should, because her chances of winning increase from one in a hundred to ninety-nine in a hundred.[40]

In our evolutionary past, small numbers mattered very much to our survival, whether we were dealing with small numbers of relatives, allies, or game animals. And small numbers could be best managed and manipulated not through probabilities but through stories with personal meaning attached. The Yanomamö's three hundred SKUs represented the universe of valuable stuff—things attached to people in the band, and thus things with stories attached. To make sense of modern economic tasks, such as grabbing a bag of chips from an aisle of store shelves, playing card games in gambling casinos, and trading stocks on Wall Street, we think in terms of stories with personal meanings, too. To pretend that it is all probabilities, that people are purely rational calculators making economic decisions, is not natural—or accurate.

For example, imagine you are a contagious disease expert at the U.S. Centers for Disease Control and you have been told that the United States is preparing for the outbreak of an unusual Asian disease that is expected to kill 600 people. Your team of experts has presented you with two programs to combat the disease:

Program A: 200 people will be saved.
Program B: There is a one-third probability that 600 people will be saved, and a two-thirds probability that no people will be saved.

If you are like the 72 percent of the subjects in an experiment that presented this scenario, you chose Program A. Now consider another set of choices for the same scenario:

Program C: 400 people will die.
Program D: There is a one-third probability that nobody will die, and
a two-thirds probability that 600 people will die.

Even though the net result of the second set of choices is precisely
the same as the first, the participants in the experiment switched preferences, from 72 percent for Program A over Program B to 78 percent for
Program D over Program C. We prefer to think in terms of how many
people we can save instead of how many people will die—the "positive
frame" is preferred over the "negative frame."[41]

The power of these *framing effects* is especially noticeable in making
decisions about investing, lending, and borrowing money. Take this financial conundrum:

1. Phones-are-Us offers the new DigiMusicCam cell phone for
   $100; five blocks away FactoryPhones has the same model half off
   for $50. Do you make the short trip to save $50?
2. Laptops-are-Us offers the new carbon fiber computer for $1,000;
   five blocks away CompuBlessing has the same model discounted
   to $950. Do you make the short trip to save $50?

Most people would take the trip in the first scenario but not the second, even though the amount saved, $50, is the same. Why? This is a
type of framing problem called *mental accounting*, where we put monies
into different categories depending on the frame, or context—in this
case, small expenditures in one accounting bucket and large ones in another. Studies show that most people are less likely to make the effort to
save money when the relative amount they are dealing with is small.

Imagine that you have purchased in advance a $100 ticket for an
event you have been eagerly anticipating, but when you arrive at the
venue you discover that you have lost the ticket and they won't let you in
unless you purchase another. Would you plunk down another hundred
bucks? In experiments in which this scenario is presented, over half of
subjects (54 percent) say that they would not. But now imagine that you
have not prepurchased a ticket to your long-awaited event, and instead
you arrive at the venue with two Franklins in your billfold, intending to

purchase your ticket at the gate. As you pull out your cash, however, you discover that one of the hundreds fell out, leaving you with just one $100 bill. Would you still purchase the ticket? Interestingly, and tellingly, a vast majority of people (88 percent) said that they would.

Rationally, there is no difference in value between a $100 ticket and a $100 bill. In economic jargon, they are *fungible*, or interchangeable. The ticket and the bill are both pieces of paper of equal value to be used as a medium of exchange. Yet emotionally there is a difference between a lost $100 ticket and a lost $100 bill. People sort their money into different categories depending on its original source (where it came from), its current status (gold, cash, product, service), and how it is spent (now or later, a sure thing or a risky gamble).

Credit cards reframe cash into a different mental accounting category that makes it much easier to spend. MIT marketing professors Drazen Prelec and Duncan Simester put this principle to the test by hosting an actual sealed-bid auction for Boston Celtics basketball game tickets. Half the people were told that if they won the bid they would have to pay for the tickets in cash, while the other half were told that if they won the bid they could pay by credit card. The cash bidders offered barely half what the credit card bidders offered.[42]

Or imagine you are offered a gamble with the prospects of a 10 percent chance to win $95 and a 90 percent chance to lose $5. Would you accept the gamble? Most people say that they would not. And yet these same people answer rather differently when the gamble is rephrased in this manner: Would you pay $5 to participate in a lottery that offers a 10 percent chance to win $100 and a 90 percent chance to win nothing? Tellingly, most of those who rejected the first gamble accepted the second. Why? Kahneman and Tversky explain it in terms of mental accounting differences: "Thinking of the $5 as a payment makes the venture more acceptable than thinking of the same amount as a loss." Why should our brains be so ill equipped to handle money? The answer, in short, is folk economics: our brains did not evolve to intuitively equate equal value commodities with their symbolic representation in paper. The logic is not the same, and so illogic prevails.

Take the following thought experiment, known as the *Wason Selection Test* (after its creator, Peter Wason, who introduced it in 1966), that is designed to test symbolic reasoning. Before you are four cards, each with a letter of the alphabet on one side and a number on the other side. Two cards are showing numbers and two cards are showing letters, such as this:

M   4   E   7

Here is the rule: *if there is a vowel on one side, there must be an even number on the other side.* Here is your task: *which two cards must be turned over in order to determine if the rule is true?* Which cards would you turn over?

Most people correctly deduce that they do not have to check the other side of card "M" because flipping over the "E" card could invalidate the rule (although by itself it does not affirm it absolutely). Most people, however, incorrectly decide to flip over the "4" card. This is wrong because the rule says that if one side is a vowel, then the other must be an even number—it says nothing about whether an even number must be accompanied by a vowel; the opposite side of the "4" card could be a vowel or a consonant. Finally, most people do not think that the "7" card must be checked, because if its flip side is a vowel then the rule is violated. Hundreds of experiments employing the Wason Selection Test reveal that fewer than 20 percent of subjects correctly deduce that the answer (in this particular example) is that the "E" and "7" cards must be inspected.

A modified version of the Wason Selection Test that personalizes the options reveals when humans do much better with logic. You are the bartender of a nightclub with a minimum legal drinking age of twenty-one. When you arrive on your shift there are four patrons at the bar. Here is the rule: *people under twenty-one cannot be served alcohol.* Here is your task: *you can ask people their age, or check to see what they are drinking, but not both.* Instead of four cards, here are your four options from which to choose:

Patron:	#1	#2	#3	#4
Drink/Age:	Water	Over 21	Beer	Under 21

In which cases should you ask patrons their age or check what they are drinking? This is a proverbial no-brainer. Obviously you do not need to check the age of #1, since the beverage is water; nor do you need to check the drink of #2, since this person is over twenty-one. Equally evident, the beer drinker, #3, could be under twenty-one, so you better check the age, and the under-twenty-one person #4 could be drinking beer, so you better check the drink. Almost everyone tested solves this logic problem correctly.

Since the logic of these two tests is the same, why do people do so poorly on the first task but so well on the second? The answer from evolutionary psychology is that the first task employs symbols and the second task involves people. In folk science terms, we did not evolve to adequately process symbolic logic problems, but as a social primate species we did evolve brain circuitry to deal with problems that involve other people, especially such social problems as deception and cheating. In response to the *Homo economicus* foundational belief that people have unbounded rationality, behavioral economists retort that we have "bounded rationality." We might say that our rationality in the modern world of symbols and abstractions is bounded by the Paleolithic environment, in which our brains evolved to handle problems in the archaic world.

In the early 1990s, citizens in both New Jersey and Pennsylvania were offered two options for their automobile insurance: a high-priced option that granted them the right to sue, and a cheaper option that restricted their right to sue. The corresponding options in each state were roughly equivalent. In New Jersey, the default option was the more expensive one—that is, if you did nothing you were automatically enrolled in that plan—while in Pennsylvania, the default option was the cheaper one. In New Jersey, 75 percent of citizens enrolled in the high-priced insurance, while in Pennsylvania, only 20 percent opted for it.

Such findings support the notion that when making decisions, we tend to opt for what we are used to, the status quo.[43] Research conducted by William Samuelson and Richard Zeckhauser reveals that when people are offered a choice among four different financial investments with varying degrees of risk, they select one based upon how risk-averse they are, and their choices range widely. But when people are told that an investment tool has been selected for them and that they then have the opportunity to switch to one of the other investments, they are far more likely to stick with the default; 47 percent stayed with what they already had, compared to the 32 percent who chose those same investment opportunities when none were presented first as the default

option.[44] The status quo represents what you already have (and must give up in order to change), versus what you *might* have once you choose.

Economist Richard Thaler calls the bias toward the status quo the *endowment effect*, and in his research he has found that owners of an item value it roughly twice as much as potential buyers of the same item. In one experiment, subjects were given a coffee mug valued at $6 and were asked what they would take for it. The average price below which they would not sell was $5.25. Another group of subjects, when asked how much they would be willing to pay for the same mug, which they did not own, gave an average price of $2.75.[45]

In our evolutionary past, this makes perfect sense. Before humans domesticated other species, they had to forage and hunt, sometimes in conditions of severe scarcity, or the threat of it, in order to survive. Those who survived most likely exhibited a strong predilection for hoarding as well. Nature endowed us with the desire to value, and dearly hold on to, what is ours. Of course, by putting so much value on what we already have, we can also overvalue it—to the point that the *sunk cost* blinds us to the value of future losses that we will sustain if we do not switch to something that we don't already have.

Cognitive dissonance drives people to rationalize irrational judgments and justify costly mistakes, whether they are end-of-the-world cultists or political leaders. Consider an especially poignant contemporary example of this *sunk-cost fallacy* that has far-reaching consequences.

The war in Iraq is now over four years old. At a cost of more than 3,100 lives plus $200 million a day, $73 billion a year, and over $300 billion since forces landed in March 2003, that's a substantial investment on the part of the United States, without accounting for the costs to other countries. American war costs are estimated to top out at over $1 trillion; and who knows how many more will die before it is done. So it is no wonder that through 2006, most members of Congress from both parties, along with President Bush and former president Clinton, believed that we had to "stay the course" and not just "cut and run." As Bush explained in a Fourth of July speech at Fort Bragg, North Carolina: "I'm not going to allow the sacrifice of . . . troops who have died in Iraq to be in vain by pulling out before the job is done." Clinton echoed the sentiment: "We've all got a stake in its succeeding."[46]

We all make similar arguments about decisions in our own lives: we hang on to losing stocks, unprofitable investments, failing businesses, and unsuccessful relationships. But why should past costs influence us? Rationally, we should just compute the odds of succeeding from this point forward, and then decide if additional investment warrants the potential payoff. But we are conditioned to overvalue the status quo.

◆ ◆ ◆

Pace Will Rogers, I am not a member of any organized political party. I am a libertarian. As a fiscal conservative and social liberal, I have never met a Republican or Democrat in whom I could not find something to like. I have close friends in both camps, in which I have observed the following: no matter the issue under discussion, both sides are equally convinced that the evidence overwhelmingly supports their positions.

This surety is called the *confirmation bias*, where we seek and find confirmatory evidence in support of already existing beliefs and ignore or reinterpret disconfirmatory evidence. According to Tufts University psychologist Raymond Nickerson, the confirmation bias "appears to be sufficiently strong and pervasive that one is led to wonder whether the bias, by itself, might account for a significant fraction of the disputes, altercations, and misunderstandings that occur among individuals, groups, and nations."[47] Experimental examples abound. In a 1981 study by psychologist Mark Snyder, subjects were asked to assess the personality of someone they were about to meet. One group of subjects was given a profile of an introvert (shy, timid, quiet), while another group of subjects was given a profile of an extrovert (sociable, talkative, outgoing). When asked to make a personality assessment, those subjects who were told that the person would be an extrovert tended to ask questions that would lead to that conclusion; the introvert group did the same in the opposite direction.[48] In a 1983 study, psychologists John Darley and Paul Gross showed subjects a videotape of a child taking a test. One group was told that the child was from a high socioeconomic class; the other group was told that the child was from a low socioeconomic class. The subjects were then asked to evaluate the academic abilities of the child based on the results of the test. The subjects who were told that the child they were evaluating was from a high socioeconomic class rated the child's abilities as above grade level; the sub-

jects who thought they were evaluating low socioeconomic kids rated them below grade level in ability. What is remarkable about this study is that the subjects were looking at the same set of test results![49]

The power of expectation cannot be overstated. In 1989, psychologists Bonnie Sherman and Ziva Kunda conducted an experiment in which they presented subjects with evidence that contradicted a belief they held deeply, and with evidence that supported that same belief. The results showed that the subjects acknowledged the validity of the confirming evidence but were skeptical of the value of the disconfirming evidence.[50] In another 1989 study, by the psychologist Deanna Kuhn, when children and young adults were exposed to evidence inconsistent with a theory they preferred, they failed to notice the contradictory evidence, or if they did recognize its existence, they tended to reinterpret it to favor their preconceived beliefs.[51] In a related study, Kuhn exposed subjects to an audio recording of an actual murder trial and discovered that instead of evaluating the evidence first and then coming to a conclusion, most subjects concocted a narrative in their mind about what happened, made a decision of guilt or innocence, then riffled through the evidence and picked out what most closely fit the story.[52]

A functional magnetic resonance imaging (fMRI) study conducted at Emory University under the direction of psychologist Drew Westen shows where in the brain the confirmation bias occurs, and how it is unconscious and driven by emotions.[53] During the run-up to the 2004 presidential election, while undergoing a brain scan, thirty men—half self-described "strong" Republicans and half "strong" Democrats—were tasked with assessing statements by both George W. Bush and John Kerry in which the candidates clearly contradicted themselves. Not surprisingly, in their assessments, Republican subjects were as critical of Kerry as Democratic subjects were of Bush, yet both let their own preferred candidate off the evaluative hook.

The neuroimaging results, however, revealed that the part of the brain most associated with reasoning—the *dorsolateral prefrontal cortex*—was quiescent. Most active were the *orbital frontal cortex*, which is involved in the processing of emotions, the *anterior cingulate*, which is associated with conflict resolution, the *posterior cingulate*, which is concerned with making judgments about moral accountability, and—once subjects had arrived at a conclusion that made them emotionally comfortable—the *ventral striatum*, which is related to reward. "We did not see any increased activation of the parts of the brain normally engaged during reasoning," Westen explained. "What we saw instead was a network of emotion circuits lighting up, including circuits hypothesized to be involved in regulating emotion, and circuits known to be involved in resolving conflicts." Interestingly, neural circuits engaged in rewarding selective behaviors were activated. "Essentially, it appears as if partisans twirl the cognitive kaleidoscope until they get the conclusions they want, and then they get massively reinforced for it, with the elimination of negative emotional states and activation of positive ones."

These neural correlates of the confirmation bias have implications that reach deeply into politics and economics. A judge or jury assessing evidence against a defendant and a CEO evaluating information about a company undergo this same cognitive process. And this brings us back to where we began: pattern-seeking, connect-the-dots thinking, fueled by the confirmation bias, that builds up incorrect folk intuitions. What can we do about it?

In science, we have built-in self-correcting machinery. Strict double-blind controls are required in experiments, in which neither the subjects nor the experimenters know the experimental conditions during the data collection phase. Results are vetted at professional conferences and in peer-reviewed journals. Research must be replicated in other labs unaffiliated with the original researcher. Disconfirming evidence, as well as contradictory interpretations of the data, must be included in a paper. Colleagues are rewarded for being skeptical. "Even with these safeguards in place," Westen cautions, "scientists are prone to confirmatory biases, particularly when reviewers and authors share similar beliefs, and studies have shown that they will judge the same methods as satisfactory or unsatisfactory depending on whether the results matched their prior beliefs." In other words, if you don't seek contradictory data against your theory or beliefs, someone else will, usually with great glee and in a public forum, for maximal humiliation.

◆　◆　◆

In their excellent 2000 book, *Why Smart People Make Big Money Mistakes*, Cornell University cognitive psychologist Thomas Gilovich and financial writer Gary Belsky consider the take-home lessons of the behavioral economics of business finance.[54] A slender, handsome man who carries about him an air of intellectual aristocracy and scientific authority, Gilovich is not only one of the most creative experimentalists working in psychology today, he is an interdisciplinary thinker who is willing to apply the research protocols from cognitive psychology to other areas of human endeavor.

An experiment run on psychology students in a university laboratory is one thing, but do people really behave that way in the real world? According to Gilovich, they do. In thinking about our tendency to overvalue sunk costs, he points to the *loss aversion* effect, which shows that people tend to fear losses about twice as much as they desire gains.

Gamblers, for example, are highly sensitive to losses, but not in the way you might think. They tend to follow a losing hand by placing bigger bets and turn conservative after a winning hand by placing smaller bets. One rationale for this strategy is "double up to catch up"—no matter how many losses in a row, if you double the bet each time, you will get back all of your money when you eventually do win. But most gamblers tend to underestimate the number and length of losing streaks. On a $10 minimum table, if you start with the minimum bet and have a losing streak of, say, eight in a row (not as uncommon as you might think), you would have to be prepared to plunk down $2,560 on the ninth hand to stick to your strategy. More important, gamblers also tend to underestimate the number and length of winning streaks and lose out on the reward of placing larger bets during them. Of course, even with an optimal betting strategy that plays to win every hand, and keeping loss aversion in check, if you play long enough you will lose because of the slight edge to the house built into the rules of the game. But casinos make even more money than the house percentage would predict because of our loss aversion.[55]

Yet the cognitive biases and fallacies that afflict us also provide useful insight into the mind of the market. Playing the stock market by buying and selling individual stocks on a regular basis is little different from gambling at a casino, and the odds are just about as good that you'll come out ahead, or at least do as well as the stock market does overall and in the long run. Studies show that even professional investors and market analysts rarely perform as well as an indexed mutual fund in the long run.

6

THE EXTINCTION OF
HOMO ECONOMICUS

Have you ever watched a white rat choose between an 8 percent and 32 percent sucrose solution by pressing one of two bars on variable-interval schedules of reinforcement? Lucky you. I devoted two years of what would otherwise have been a misspent youth to running choice experiments with rats in the psychology department at California State University, Fullerton, for a master's thesis entitled "Choice in Rats as a Function of Reinforcer Intensity and Quality."[1] Boys gone wild!

I was investigating the *matching law*, discovered in 1961 by the Harvard psychologist Richard Herrnstein (he of the later *Bell Curve* controversy over black-white I.Q. score differences), which states that organisms will match their rate of responding to the rate of reinforcement. In a two-alternative choice in which one variable-interval (VI) schedule pays off, on average, once a minute, while the other VI schedule pays off, on average, once every three minutes, the organism will emit three times as many responses on the first schedule as on the second.[2] "Organism," by the way, denotes any animal, although ever since B. F. Skinner's 1938 book on *The Behavior of Organisms* founded the field of behaviorism, "organism" has usually meant rats, pigeons, and introductory psychology students. Behaviorism is based on the belief that anything any organism does, whether it is in the form of observable actions and movements or internal thoughts and emotions, should just be considered "behaviors" without conjecture about what is going on in the mind. And in this context, the matching law did not make sense: If the first schedule pays three to one over the second, why bother responding to the second schedule at all? Why not just press the bar that pays off the most?

But if we constantly choose the same option over and over, it grows less appealing. Like a ring we fail to notice on our finger, our senses habituate to the same stimulation. To break the psychology of *habituation*, we have to break the pattern of choice. Switching choices resets the brain to once again notice the previously habituated stimulation, thereby pushing the organism to match the frequency of responses to the intensity of the enjoyment they produce.[3] Steak is great, but not every night. A little pasta now and then cleanses the palate and resets the bar for steak's desire.

Choices in the real world, however, are always more complex than those in the laboratory. In my two years with the rats, I wanted to know if the matching law also applied to reinforcer intensity and quality. In experiment one, I tested reinforcer intensity by giving my rats a choice between 8 percent and 32 percent sucrose on a standard VI schedule, and in experiment two, I tested reinforcer quality by giving them a choice between 8 percent sucrose and 8 percent sucrose plus 4 percent salt on the same VI schedule. In keeping with the tradition of scientific objectivity, my rats were numbered 1 through 8, but I confess that I also named them after the Los Angeles Dodgers baseball team's starting lineup that year. For some reason, little Dusty Baker was especially unruly. If you put the eraser end of a pencil inside his cage, he would promptly shred it.

In both of my experiments I found an undermatching effect, in which my rats responded less to the schedule with the higher rate of reinforcement than predicted by the matching law when reinforcement intensity or quality was added to the choice: that is, the more variables added to a choice, the more complicated the decision and the less predictable the behavior. That was not an especially surprising finding, given the fact that whenever you add layers of complexity to an environment it compounds the choices we make. Yet I did wonder what was going on inside the little brains of my rats as they made their decisions on which bar to press and how many times to press it. And it made me wonder what is going on inside the big brains of humans when we make decisions about the equivalent real-world choices. For behaviorists of that time, in the late 1970s, this was a nonquestion. The brain was a black box, impenetrable by the tools of science, and the mind was equivalent to Churchill's Soviet Union: a riddle wrapped in a mystery inside an enigma.

Thanks to cognitive psychologists and neuroscientists, the black box has now been opened. Although many brain riddles remain, the enigma that is the mind is becoming a whole lot less mysterious.

◆ ◆ ◆

Imagine that your child's private school tuition bill of $20,000 is due and the only source you have for paying it is selling some of your stock holdings. Fortunately, you invested in Apple before the iPod boom, purchasing 400 shares at $50 each, for a total investment of $20,000. The stock is now at $100 a share. Should you realize your net gain by selling half of your Apple stock and use the proceeds to pay off your bill? Or should you sell off that loser Ford stock you purchased ages ago for $40,000, at its current value of $20,000?

As we began to see in the last chapter, most people (me included) would sell the Apple stock and hang on to the Ford stock in hopes of recovering their losses. Financially, this would be the wrong strategy (unless you prefer to factor in the stock loss to your future tax returns). Why would you sell shares in a company whose stock is on the rise and hang on to shares in a company whose stock is in the dumpster?

Consider a slightly more complicated thought experiment that reveals another irrationality of the standard of *Homo economicus*. Imagine that I gave you $100 and a choice between (A) a guaranteed gain of $50 and (B) a coin flip in which heads gets you another $100 and tails gets you nothing. Do you want A or B? Now imagine that I gave you $200 and a choice between (A) a guaranteed loss of $50 and (B) a coin flip in which heads guarantees you lose $100 and with tails you lose nothing. Do you want A or B? The final outcome for both options A and B in both scenarios is the same—their utility, in neoclassical economic terms, is equal—so rationally it does not matter which option you choose; therefore people should choose both equally. According to Rational Choice Theory, which forms the foundation of *Homo economicus*, we maximize our utility when we make decisions. That is, when faced with a choice, we consider the value of the outcome and make a rational decision about the most efficient course to take to get to that end.

But human loss aversion also shapes our choices. Most people choose A in the first scenario (a sure gain of $50) and B in the second scenario (an even chance to lose $100 or nothing). We are willing to be bold when faced with the possibility of losses and are cautious when granted the chance to lock in a sure gain.[4] Emotionally speaking, there is a difference—a big difference—between the two choices, even though there is no difference between having $100 and a sure or potential gain of $50 and having $200 and a sure or potential loss of $50.

Thousands of experiments in behavioral economics demonstrate conclusively that most of us are highly *risk averse*. And the research shows precisely just how averse to risk we are—as we saw in chapter 5, on average most people will reject a fifty-fifty probability of gaining or losing money, unless the amount to be gained is at least double the amount to be lost. That is, most of us will accept a fifty-fifty chance to gain $100 if the risk of losing money does not exceed $50.[5] When the potential payoff is more than double the potential loss, most of us will take the gamble. And science can now explain the difference.

To wrap my mind around the incredibly rich field of behavioral economics, I audited a course taught at the California Institute of Technology (Caltech) by Colin Camerer, an experimental economist who caught the behavioral wave early. With pale blue eyes and a husky build, Camerer is the consummate careerist one has to be to succeed at Caltech. A child prodigy, Camerer received his B.A. in quantitative studies from Johns Hopkins University at age eighteen and was a newly minted Ph.D. in behavioral decision theory from the University of Chicago at age twenty-two. Punctuating his speech with flashing eyebrows, Camerer is a rapid-blinking, fast-talking lecturer who occasionally starts the next sentence before the previous one has ended. He engages his students with humor and pop culture references (Britney Spears and

Lindsay Lohan appear in opening-night examples), and since he's now a new parent, he can't seem to help referencing the behavioral economic choices made by his one-year-old, which break up the technical jargon and mathematical equations.

"The brain is the organ of economic decision," he begins. "With behavioral economics we want to use facts and constructs to reveal the limitations on computation, willpower, and self-interest"—a direct attack on the presumption of *Homo economicus*. "In fact," he tells me when I visit his office, "the first time I heard about rational choice theory I couldn't keep a straight face. There are just so many exceptions to the rule that it made me think that maybe we need a new rule."

Much of behavioral economics involves challenging the concepts of the Nash equilibrium and the Pareto optimum. Recall that in a Nash equilibrium, two or more players reach an equilibrium where neither one has anything to gain by unilaterally changing strategies. Applied more generally, markets reach a point of equilibrium where holding on to strategies is more profitable than switching strategies, and this leads to market stability.[6] Pareto optimum is reached when both parties gain, or one party gains, but the other does not lose. Applied more generally, markets reach a point of optimization when no further trades can be made without someone losing.[7] The problem with equilibrium theory is that it makes certain assumptions—perfect competition, perfect information, perfect rationality—that behavioral economists have discovered (and most lay people have long known) are not true. Competition is not perfect, people do not have complete information, and we do not make rational choices. What behavioral economists like Camerer want to know is how human irrationality in markets alters market balance.

"In general, rational choice theory still gets it right often enough that it provides a useful approximation of how the world works," Camerer says. "For example, people really do respond to incentives in the long run. But there is more to the story." Camerer wanted to do "economics from the inside," meaning inside the head. In fact, Camerer's first job was in the business school at Northwestern University, where he taught a course on why businesses succeed or fail. From his studies he realized that it often comes down to the personalities of the people running the company, the psychology of how they make decisions, how they treat their employees, and how they engage the marketplace. To understand that, you need to understand human psychology, which led Camerer to the work of Daniel Kahneman and Amos Tversky.

Psychologizing about what is going on inside the head is one thing, but Camerer wants to quantify it in the lab, constructing mathematical predictions based on these lab results and then testing the principles in the real world. For example, Camerer studied New York City taxicab drivers who pay a fixed fee to rent their cabs for twelve hours and keep all revenues earned in that time. Camerer found that when cabbies set for themselves an average they had to make each day in order to pay for the cab, their aversion to loss led them to work longer hours on rainy days in order to make their goal and to quit early on good days as soon as their goal was met.[8]

In the classroom, Camerer reveals a preference for mathematical precision in studying human choice in markets. "People act as if they have a utility function and they make choices toward maximizing their utility," he explains, dashing off equations and logarithmic curves on the white board that stands next to his laptop. Take something simple, like the *law of supply and demand*, which predicts that if the price of a good is at a low enough level to cause consumers to demand more of it than producers are prepared to supply, the price will go up until demand decreases. Conversely, if the price of a good is at a high enough level to cause consumers to demand less of it, producers will drop the price until it reaches a level where consumer demand will increase, which will then cause the price to increase. Market equilibrium is reached at the point where the quantity supplied is approximately equal to the quantity demanded, and the balance is maintained through this interaction of consumers, producers, and prices. It is another example of an autocatalytic feedback loop, a self-driving system that economists like to think is as lawlike in its predictability as equivalent phenomena in physics (weather systems) and biology (ecosystems).

So powerful and pervasive is the law of supply and demand that there may even be a deep evolutionary foundation to it. A 2006 study by the

Yale University economist Keith Chen and his colleagues, for example, found that the law holds up in trading with capuchin monkeys. In their experiment, the monkeys were given twelve tokens that they were allowed to trade with the experimenters for either apple slices or grapes, both of which they like equally well (unlike cucumber slices). In one trial, the monkeys were given the opportunity to trade tokens with one experimenter for a grape, and in another experiment for apple slices. The capuchins are motivated to trade because the food they receive for the tokens is part of their daily caloric intake on the days that the experiment is run—in other words, they're hungry! One capuchin monkey in the experiment, for example, traded seven tokens for grapes and five tokens for apple slices. A baseline like this was established for each monkey so that the scientists knew each monkey's preferences. The experimenters then changed the conditions. In a second trial, the monkeys were given additional tokens to trade for food, only to discover that the price of one of the food items had doubled. According to the law of supply and demand, the monkeys should now purchase more of the relatively cheap food and less of the relatively expensive food, and that is precisely what the monkeys did. Further, in another trial in which the experimental conditions were manipulated in such a way that the monkeys had a choice of a 50 percent chance of a bonus or a 50 percent chance of a loss, as predicted by the principle of loss aversion, the monkeys were twice as averse to the loss as they were motivated by the gain.[9]

Remarkable! Monkeys show the same sensitivity to changes in supply, demand, and prices as people do, as well as displaying one of the most powerful elements in all of human behavior, loss aversion. It is extremely unlikely that this common trait would have evolved independently and in parallel between multiple primate species at different times and different places around the world. This suggests that there is an early evolutionary origin for such preferences and biases, and that the trait evolved in a common ancestor to monkeys, apes, and humans and was then passed down through the generations—in this case to the lineage that led to capuchin monkeys and to the lineage that led to us. Thus, if there are behavioral analogies between humans and other primates, the underlying brain mechanism driving the choice preferences most certainly dates back to a common ancestor over ten million years ago.

Ten million years ago, the psychology behind the law of supply and demand evolved in the earliest primate traders.

✦ ✦ ✦

You have probably seen those ads comparing incandescent lightbulbs with compact fluorescent lightbulbs, the latter of which are being touted as environmentally sound because they provide the same amount of light but last ten times longer. Energy efficiency means using less electricity, which means burning less coal, which translates into lower carbon emissions into the atmosphere, which will help solve the problem of global warming. That's the argument. But if you are a global warming skeptic, or if you base your purchases on more quotidian criteria such as your personal bottom line, the ads slant toward selling you on the long-term cost savings. That is, they are appealing not to your environmental consciousness, but to your preference for personal profits. In just one example among many (the calculations vary but the point remains the same), a 60-watt conventional bulb costs $0.75 and lasts one thousand hours, compared to a 14-watt compact fluorescent lamp that costs $2.00 and lasts ten thousand hours. You get the same amount of light from both. The total cost to operate each bulb for ten thousand hours, including the bulb replacement cost, is $59 for the incandescent bulb and $12 for the compact fluorescent lamp. Which one should you buy? It depends on your personal time preference. Do you want an immediate reward (you can save $1.25 right now), or a future payoff (you can save $47 down the road)?

The delay of gratification and *time preference*—how we discount value over time—is a hot topic in behavioral economics. The lightbulb example is what is known as an *intertemporal choice*—decisions that involve tradeoffs among costs and benefits occurring at different times—and research shows that in order to get people to prefer long-term options you have to give them an incentive. That is, we need to be paid to delay gratification. In a 1981 study, the economist Richard Thaler gave subjects a choice between taking $15 today and some higher amount in the future, and then asked them to tell him what that higher figure would have to be if the future date were in one month, one year, or ten years away. The median responses were $20 in one month, $50 in one year, and $100 in ten years.[10] In economic jargon, the utility of the original offer diminishes over time, so the future value of it must be increased today.

Ever since Thaler's findings, hundreds of such experiments have been conducted, revealing many subtle differences but all demonstrating the same fundamental principle that time discounts value. Here's another example. Which would you rather have: $20 today or $22 in one week? Most people take the twenty bucks today because it isn't worth waiting an entire week just for two bucks. But let's change the time dimension and ask which would you rather have: $20 in seven weeks or $22 in eight weeks? Most people opt to wait the extra week. That's weird. In both cases, you are waiting a week to get an extra two dollars. The explanation is in the difference between going from 0 to 1 and going from 7 to 8. Compared to now, a week is a long time. Compared to having already waited seven weeks, delaying gratification one more week is trivial.

Behavioral economists have developed elaborate (and intimidating) mathematical equations to try to understand and predict these intertemporal choices and have uncovered some variables that influence our willingness to delay gratification. Some are fairly obvious, such as whether the item in question is visually available—if it is, you are more likely to want it now, while if it is hidden from view, you are more likely to forgo immediate gratification. That's why if you are trying to lose weight, you should not even look at the dessert menu. Of course, if you are a restaurant server trying to increase the value of the bill, your best bet is to bring the dessert tray to the table and ask your customers which one they would like. As my father always told his salesmen at his Ford dealership, never ask the customer *if* they want to buy a car; walk them around the lot and ask them *which* car they want to buy.

A still deeper question is why the time effect exists at all. The answer most likely comes back to our evolutionary ancestors. There are uncertainties about the future that increase the further out you project. Who knows what might happen to my money in a week, so I better take what I can get now and not risk losing it all later. In studies on the delay of gratification across species, there is a steady and predictable increase in the ability to put off into the future rewards that could be had today as the size of the cerebral cortex increases. Rats and pigeons have an extremely short delay of gratification; dogs a little bit longer; primates longer still. But for all of these species the delay of gratification is measured in seconds or minutes, whereas humans can delay gratification for years.[11] Being able to delay gratification requires that rational cognition be capable of overriding emotional impulsivity, and you need a lot of cortex for that. Further, studies of patients with brain damage from strokes, accidents, or surgery clearly indicate that damage to the prefrontal cortex usually leads to impulsivity and an inability to plan for the future.[12]

With the advent of functional MRI scans of the brain, we can pinpoint with even greater precision where impulsivity and delay of gratification happen. In a study by the Princeton University neuroscientist Samuel McClure and his colleagues, people were offered a choice between differing monetary rewards (between $5 and $40) available at different points in time (today, in two weeks, in one month) while inside the fMRI scanner. The scientists made three significant findings: (1) for immediately available rewards, there was a significant increase in activity in the *limbic system* associated with the midbrain dopamine system— well known to be involved in drug addiction and impulsive behavior; (2) regardless of the delay in payoff, there was a significant increase in activity in the *lateral prefrontal cortex* and *posterior parietal cortex*— implicated in higher-level deliberative processes and cognitive control, and probably activated during the consideration of economic options; and (3) when subjects chose the longer-term options, only their *frontoparietal lobes* lit up—this area is associated with the ability to value greater rewards even if they are delayed. In other words, the brain deliberates over both short-term and long-term rewards, but if the "low-road" emotional limbic system kicks into high gear it can override the "highroad" rational prefrontal cortex, unless the still higher superrational frontoparietal cortex overrides the entire system. The researchers concluded from these findings "that human behavior is often governed by a competition between lower level, automatic processes that may reflect evolutionary adaptations to particular environments, and the more recently evolved, uniquely human capacity for abstract, domain-general reasoning and future planning."[13]

Not only is there a linear march in cortex-size increase from simpler to more complex mammals, there is a temporal march in cortex-size increase from ancient to modern mammals through evolutionary history. The low-road brain structures, available in all mammals and primates, give us all the default decision of impulsivity and immediate gratification. But as higher-road neural networks evolved along with the development of the cerebral cortex, particularly in the higher primates, the capacity to override the limbic impulses to act now also evolved, and our Paleolithic ancestors learned to discount the future or delay gratification. However, most trade in our Paleolithic past was almost exclusively in basic commodities such as food, clothing, and social favors. The informal means of conflict resolution and trust enforcement were not always reliable, and there were no social institutions in place to create long-term trust and enforce future payoffs and expected reciprocity. So the future discounting effect is ultimately due to the uncertainty of Paleolithic futures.

✦ ✦ ✦

To learn more about the neuroscience behind economic choices, I visited the lab of neuroscientist Russell Poldrack and behavioral economist Craig Fox at UCLA and climbed inside the claustrophobically cramped quarters of the magnetic tube that is the MRI machine. The factory showroom model MRI weighs in at around twelve tons and invoices at about $2.5 million, which does not include installation, training, and maintenance, which can drive the bill up another million. This is big science employing big machines with big price tags that require big budgets.

The technology behind the MRI, and the computer-generated statistical methodologies employed to convert raw MRI images into colorful brain pictures, is every bit as impressive as the price tag. Loosely speaking, neural activity in the brain is highly correlated with changes in blood flow and blood oxygenation. When neurons are active, they consume more oxygen, pulled out of the hemoglobin in red blood cells from nearby capillaries; the brain responds to this increased need for oxygen by sending more, and (for reasons that are not yet fully understood) actually sends more than is needed. There is a short delay of about five seconds between neural activity and blood flow change, which leads to

differences in the relative concentration of oxygenated hemoglobin in those neurally active brain areas. Because hemoglobin is magnetically sensitive, there are measurable magnetic differences between oxygenated and deoxygenated blood, and these magnetic differences are what the fMRI is measuring. How does the MRI measure these magnetic differences, and how do those differences get converted into those colorful brain scan pictures we've all seen?

As Poldrack walked me through the lab, he talked me through the process. The MRI scanner is a large electromagnetic cylinder constructed of superconducting wire cooled by helium, and in accordance with Maxwell's equations describing the entangled relationship between electricity and magnetism (as when huge spinning magnetic turbines inside dams generate electricity as escaping water turns them), this creates a magnetic field inside the tube. Measured in Tesla (T) units (after the physicist and inventor Nicola Tesla), with 1 T the equivalent of twenty thousand times the earth's magnetic field, most MRI scanners put out 1.5 to 4 T. This is magnetism on steroids! So powerful is the MRI magnetic field that subjects must be stripped of all metal objects before entering the room (people have been killed by flying metal objects in an MRI), and people with pacemakers or metal implants cannot even go in the room, which itself is heavily shielded in steel and features soundproofing technologies to muffle the bone-shaking noise produced when the magnets work their magic. And magic it surely seems, for the theoretical physics upon which the technology is based sounds positively legerdemain.

When a person is placed inside this magnetic field, some of the atoms in the body's tissues align to the magnetic field. It's only about one in a million atoms that are so aligned, but because there are about 7 octillion (seven thousand quadrillion, or seven thousand thousand trillion) atoms in the body, this works out to about six million billion atoms in a $2 \times 2 \times 5$-millimeter cube of tissue—plenty for the scanner to read. The protons in the nucleus of each atom are spinning, and like a spinning top they also precess (or wobble, as the axis of rotation sweeps out a cone). The frequency at which a proton precesses depends upon the strength of the magnetic field—which varies along the length of the tube—and this "gradient" is slightly higher at the head end, causing the protons to

precess at a slightly different frequency. In order to make an image, electromagnetic energy is sent in at a particular radio frequency, which excites the protons to match that resonant frequency caused by the magnetic field. This, in effect, tips the direction of their alignment to the side. Over time (milliseconds), these protons come back into alignment with the main magnetic field, and in the process they shed some energy. It is this energy that is being measured by the head coil (affectionately called the "cage") in which the subject's head is placed, and this is what is used to create the image. Inside the cage, signals are registered and encoded from different parts of the head by measuring differences in frequency and phase that result from the magnetic field changing directions—left and right, front and back—and it is these shifts in the magnetic field that generate the MRI's earsplitting noise.

Once the subject is jammed into the tube, with head locked firmly in place to reduce head motion (which can blur the images), the experiment begins. The MRI scanner snaps a picture of the brain every two seconds while the subject watches images, pushes buttons, or makes decisions, all of which is presented through goggles featuring tiny screens representing the screen of the computer on which the experiment is run. Corrections for head motion are made by lining up all of the individual two-second images with one another, and then the data from all subjects are warped together to correct for differences in brain size and shape. The computer then blurs the data to help reduce small differences in the location of structures within different subjects' brains. A statistical model is then generated to show how the MRI signal should change over time in an area that responded perfectly to the task, followed by corrections on the statistical maps to account for the fact that there is such a large number of statistical tests being conducted. Finally, additional statistical tests are run to compare the observed data with the perfect model, resulting in statistical maps that are then converted into colorful pictures of brains in action. Keep all this in mind the next time you see one of those colorful brain scan photographs in a popular magazine with an arrow pointing to some spot that says "This is your brain on X." That image does not represent any one person's brain, but is instead a statistical computation of the entire subject pool rendered with artificial colors to highlight the differences.[14]

Poldrack, Fox, and their UCLA colleagues Sabrina Tom and Christopher Trepel employed this technology in the service of studying the neural basis of loss aversion while subjects made decisions under risk. In the experiment, subjects were presented with the prospect of accepting or rejecting a gamble that offered a fifty-fifty chance of gaining or losing money. As the potential for gains rose they found increased activity in a set of brain regions that receive input from the *mesolimbic* and *mesocortical* dopamine systems (dopamine is a neurotransmitter substance associated with motivation and reward). As the potential for losses increased, they found decreasing activity in these same reward-sensitive areas. Interestingly, it appears that both losses and gains are coded by the same brain structures—the *ventromedial prefrontal cortex*, associated with decision making and learning in the context of rewards and punishment, and the *ventral striatum*, associated with learning, motivation, and reward. Individual differences in loss aversion (i.e., how much more sensitive to losses versus gains the person was in making a decision) were predicted by how much more the brain was turned off by losses versus turned on by gains. This effect may be caused by differences in neurochemistry, which in turn may be caused by genetic variation. That is, some of us may be hardwired to be high or low risk takers, and this translates into real-world financial prospects, good and bad.[15]

What are the implications of this research for why people make money mistakes? I put the question, and others, to Poldrack and Fox.[16] Attired in the laid-back style of a busy academic scientist—khaki slacks, untucked short-sleeved polo shirt, studious black-rimmed glasses—Poldrack is bookishly smart but modestly understated in his demeanor. He's much deeper than first appearances let on, but with a little probing, the nimble mind of a wickedly smart brain scientist emerges. Trained in both psychology and neuroscience, Poldrack is more broadly interested "in developing experiments that can help test theories about how the brain makes decisions; that is, what computations are performed by individual brain regions or networks, and how those computations are used in service of decisions." But is the brain a collection of specialized modules—analogous to a Swiss Army knife—or is there a more general operating system running the show—analogous to Microsoft Windows Vista or Macintosh OS X?

Poldrack admitted that interpreting fMRI scans is as much an art as a science. "It is tempting to look at one of those spots and say 'this is where X happens in your brain,' when in fact that area could be lighting up when involved in all sorts of tasks," he said. "Take the right prefrontal cortex that lights up when you do almost any difficult task. One way to think about it is in terms of networks, not modules. When you are engaged in thinking about money, there's a network of several different areas involved in communicating with each other in a particular way. So, the prefrontal cortex may be involved in a lot of different tasks, but in communication with specific other brain networks it becomes active when engaged in one particular task, such as thinking about money." Teasing these apart requires making relative comparisons across a variety of tasks. Choice experiments work especially well with fMRI, because decisions provide contrasts between tasks, giving the neuroscientist something to compare.

It is easy to succumb to the temptation to oversimplify the modularity of the brain, as in, *here* is the module for reason, *there* is the module for emotion, or *this* neural pathway represents the emotion low road and *that* neural pathway represents the reason high road. It turns out the neural networks of the human brain are quite complex and interconnected. "There *are* rational and emotional ways of thinking," Poldrack assures me, but "it turns out that they interact with one another a lot."

"On the other hand," Craig Fox chimes in, "in order to make progress in science we need to simplify and abstract, especially when studying complex systems." Fox is a study in contrast with Poldrack. His apparel more closely resembles that of the Fortune 500 company execs to whom he has lectured on the psychology of corporate mergers and why they so often fail (primarily as a result of the host of cognitive biases discussed in the last chapter). Fox studies how people make judgments and decisions under uncertainty, using a wide range of populations from options traders to executives to lawyers to sports fans, and using a wide range of experimental methods from surveys to laboratory and field experiments to analysis of archival data to brain imaging studies (with Poldrack). Having trained under the founding fathers of the field—Kahneman and Tversky—Fox has hopes for the technology that are more directly economic than Poldrack's.

"I think of fMRI as a very promising new tool for studying decision making under risk, with a unique set of strengths and weaknesses, that complements traditional tools such as lab experiments and analysis of field data."

As an example, Fox noted that in behavioral economics experiments, Kahneman distinguished two different types of utility: the value you experience while you are making a decision and the value you experience after the decision is made. One experimental condition entailed putting a subject's hand in an ice bath at 14 degrees Celsius for sixty seconds, and the other condition was at 14 degrees Celsius for sixty seconds followed by a gradual warming to 15 degrees Celsius over the next thirty seconds. The result was startling, Fox noted. "Most participants rated the ninety-second experience as *less* unpleasant than the sixty-second experience, and most said they would rather repeat the ninety-second experience than repeat the sixty-second experience." In other words, participants in the experiment actually preferred *more* pain to less! In the language of behavioral economics, the *experienced utility* (the pain experienced moment by moment) can be very different from the *retrospective utility* (the recollection of the aggregate pleasantness or unpleasantness of the experience). This is known as the *peak-end rule*, in which we judge a past event almost entirely on how the experience was at its peak and at its end—whether pleasurable or painful—instead of a net average for the entire duration of the event.[17] Controlling for these two types of utility, in their fMRI experiments Poldrack and Fox identified the brain network associated with such evaluations. "The surprise finding in our study was the fact that it was a single brain network processing both gains and losses in the gamble, and that this brain network turned down at a faster rate to losses than it turned up to gains. It is like there is a neural value function."

When economists talk about "value function," they usually mean something technical and specific, and it is often expressed in a mathematical equation, but it is enough to know that the value function is an estimate of the expected future reward, or value, of a purchase. More generally, value is equated with utility, which is a throwback to the nineteenth-century doctrine of utilitarianism, Jeremy Bentham and John Stuart Mill's notion that social and political policy should be aimed at maximizing the total utility of as many people as possible in society, or "the greatest happiness for the greatest number." In economics, utility (or value) is a measure of the relative happiness or satisfaction gained from a purchase.

If utility and value bring happiness and satisfaction, and happiness and satisfaction are emotions associated with brain activity, then it stands to reason that there must be areas of the brain associated with utility and value. Sure enough, there are, and not just in people. For example, in 1999, New York University neuroscientists Michael Platt and Paul Glimcher gave monkeys a choice of looking in different directions, one of which was rewarded with a squirt of juice that the monkeys enjoyed, while the scientists recorded the activity of single neurons in an area of the brain called the *lateral intraparietal*, or LIP (located just above the ears and toward the back of the brain), associated with attention, decision making, and preparation for making motor movements. Platt and Glimcher found a linear relationship between the rate of neural firing and the value of the reward (juice). In their words: "Our data indicate that some of the variation in the LIP activity was, in fact, correlated with economic decision variables and that these variables influenced the choices made by our animal subjects and neural activity in a similar manner."[18] That is, as the utility of the reward increased, so too did the rate of firing of the neurons that process such information. It appears that the brain really is computing utilities.

Numerous studies on primates measuring single neurons and on humans using fMRI brain scans have confirmed that there are neural networks associated with computing value and utility, which themselves are directly associated with behavior in a market decision. For example, according to research on prospect theory, people are on average about twice as sensitive to losses as they are to gains, but they vary widely around this average. Do variations in people's brains predict such variations in their behavior? Amazingly, they do. Fox and Poldrack found that the correlation between the behavior of their subjects with the neural activity in their brain was 0.85.

This finding is more profound than the technical verbiage describing it might intimate. The correlation coefficient is one of the most common statistics used by social scientists, and it is represented by r, which has a range from 0.00 to 1.00—from no relationship to a perfect relationship.

For example, the correlation between height and weight is $r = 0.70$. If you square this number it gives us what we call the *coefficient of determination*, or $r^2 = 0.49$. That is, 49 percent of one's weight is directly accounted for by one's height, and vice versa. In most social science research, correlations fall well below $r = 0.50$, even in studies with identical twins. For example, the correlation of religious interests between identical twins raised apart is $r = 0.49$, a very significant figure that indicates a strong genetic component to religiosity. And yet this correlation yields an $r^2 = 0.24$, meaning that only 24 percent of the variation between people on their religiosity can be accounted for by their genes; the rest is determined by environment. (So much for the accusation of genetic determinism to the findings from twin studies.) Thus, the correlation between behavior and brain of $r = 0.85$ yields an $r^2 = 0.72$, which means that 72 percent of the variation in subjects' decisions in this experiment can be accounted for by brain activity. This was a startling finding for psychologists, accustomed to dealing with an r^2 in the low two-digit or high one-digit figures.

✦ ✦ ✦

There are few truisms in economics truer than the fact that there is no free lunch. If you want something, you have to pay for it. The something you want has a positive value, but the paying for it has a negative value. Value-seeking shoppers, of course, want to maximize the former and minimize the latter. In fact, at the core of all economic theory is its ability to explain and predict the decision someone makes as to whether or not to purchase a product. Integrating behavioral economics and neuroeconomics, Stanford University neuroscientist Brian Knutson and his colleagues employed fMRI scans on subjects presented with a decision whether or not to purchase a product. Knutson wanted to know where in the brain the shopping module is located.

Utilizing an experimental task called "Save Holdings or Purchase"— SHOP for short—twenty-six subjects had their brains scanned while they decided whether or not they wanted to purchase a box of Godiva chocolates or a DVD of the popular television series *The Simpsons*. Four seconds after each item was presented, a price was offered below the item. After another four seconds, the subject was presented with two boxes that appeared on either side of the screen, one with "Yes" in it and the other with "No." He or she then made a decision on whether or not to purchase the product. There were eighty trials total. To avoid left-right behavior preferences, the "Yes" and "No" boxes were randomly switched. Some of the purchases were virtual, but others were actual, in which the subject received $20 and was given the opportunity to actually purchase the Godiva chocolates or *Simpsons* DVD.

So what did Knutson and his colleagues discover about the brain's activity during shopping (or at least during shopping inside an fMRI brain scanner)? When a product appears, it activates the *nucleus accumbens* (*NAcc*), a network of neurons near the middle of the brain associated with the reward center and involved in appraising the item. When the price of that product appears, it activates the *mesial prefrontal cortex*, a region of the brain affiliated with higher executive functions and decision making. In varying the experimental conditions, the researchers found that the activity of the mesial prefrontal cortex varied according to the difference between what someone would pay for a product and what it actually cost. Finally, the *insula* lit up, which is an area of the cortex well known to respond to negative stimuli, such as foul smells, becoming more active when people decided not to make the purchase. That is, they apparently experienced a negative emotion, almost like disgust, when the price was perceived to be too high. As the researchers concluded: "The findings are consistent with the hypothesis that the brain frames preference as a potential benefit and price as a potential cost, and lend credence to the notion that consumer purchasing reflects an anticipatory combination of preference and price considerations."[19] The shopping choice produced an emotional reaction, not a rational decision. Knutson explained, "What we're looking at is not so much the brain's reaction to products and prices as a person's subjective reaction to the products and prices. Is the product preferable? And is the price too much?" It is possible, Knutson suggested, that such neuroeconomic research could help illuminate brain differences between using cash and using credit cards. "Maybe a shopper's insula isn't engaged when they're using credit cards," he speculated. "This is the first study to show that you can use brain activation alone to predict purchasing on a trial-by-trial basis."[20] But if higher prices activate the disgust network in the brain, then do lower prices, great deals, and successful investments activate the pleasure center in the brain? To find out, Knutson conducted another fMRI study, this one on nineteen subjects ranging in age from twenty-four to thirty-nine, each of whom was given $20 to invest in one of three investment vehicles: (1) a bond that guaranteed a return of $1 per round, (2) a safe stock that offered a 50 percent chance to earn $10 per round but a 25 percent chance of losing $10 per round, and (3) a risky stock that offered a 50 percent chance of losing $10 and a 25 percent chance of winning $10. Subjects did not know ahead of time what each stock would pay off, so they had to base their decisions on the history of the stock.

The results were revealing. Three-quarters of the time, subjects made rational investments, but one-quarter of the time they made bad investments, picking one of the stock options when they should have chosen the more conservative bond. Whenever subjects made a risky investment by choosing one of the stock options that they had lost on before, what lit up most strikingly was the peanut-sized nucleus accumbens, which we just saw is associated with reward and pleasure. In fact, it is worth noting that the NAcc appears to be associated with the so-called pleasure center of the brain that is fueled by dopamine and has been implicated in the high from cocaine and orgasms. In 1954, James Olds and Peter Milner accidentally implanted an electrode into the NAcc of a rat and discovered that it became very energized, so they purposely set up an apparatus such that whenever a rat pressed a bar it generated a small electrical stimulation to the area. The rats pressed the bar until they collapsed, even to the point of forgoing food and water.[21] Now that is the very definition of utility maximization! The effect has since been found in all mammals tested, including humans—people who have undergone brain surgery and had their nucleus accumbens stimulated equated the feeling with an orgasm.[22]

In Knutson's experiment, he found that whenever his subjects made a safe investment, their anterior insula became active. We saw earlier that the insula is associated with disgust and the processing of negative emotions, so it is possible that the anterior insula may provide a brake on making risky investments. If the NAcc reward center is fueled by dopamine, this nonreward center is energized by serotonin and norepinephrine (also known as noradrenaline, and long known to be associated with fear and anxiety). Knutson's research suggests that when we are faced with a risky decision there is a tension between these different and sometimes competing brain regions. He found that subjects whose brains showed a lot of activity in their anterior insula were 20 percent less likely to invest in a risky stock, even when it appeared that the stock might finally pay off. "These findings are consistent with the hypothesis that NAcc represents gain prediction, while anterior insula represents loss prediction," Knutson concluded.

Brain modules for winning and losing? Who knew? Casino owners, that's who. "This may explain why casinos surround their guests with reward cues," Knutson explained, such as cheap food, free booze, discounted room rates, and surprise gifts, because "anticipation of rewards activates the NAcc, which may lead to an increase in the likelihood of individuals switching from risk-averse to risk-seeking behavior. A similar story in reverse may apply to the marketing strategies employed by insurance companies."[23]

So sex and shopping are not so far apart, particularly in the brain, and we would be well advised to listen to the whisperings from within of what those feelings are trying to tell us about the possible consequences of our actions.

7

THE VALUE OF VIRTUE

You are walking along a railroad line when you come upon a fork in the track and a switch. There are five workers on one track and one worker on the other track. Suddenly, you realize that a trolley car is hurtling along and is about to hit and kill the five workers unless you throw the switch and divert the car down the other branch, killing the one worker instead. Kill one to save five. Would you throw the switch? Most people say that they would. In a second scenario, instead of coming upon a switch, you happen across a bridge where there is a large man standing next to you. The trolley is once again speeding down the track and is about to hit and kill the five workers, unless you push the large man onto the track, killing him but stopping the car. Kill one to save five. Would you throw the man? Most people say that they would not.[1] Since the moral calculation is the same, logically it should not matter. But emotionally it does. Why?

The reason is that switches and people are categorically different, and evolutionary theory explains why. Evolution designed us to value humans over nonhumans, kin over nonkin, friends over strangers, in-group members over out-group members, and direct action over indirect action, because these differences impacted survival and reproduction. These

intuitively felt differences and moral intuitions reflect a rational calculation conducted over the evolutionary eons. What may seem like irrational
behavior today may actually have been rational deep in our Paleolithic
past. Without an evolutionary perspective, the assumptions of "economic
man" as selfish, rational, and free make no sense. The reason has to do
with the evolution of our *moral emotions*.

◆   ◆   ◆

In 2000, my mother fell and hit her head on the corner of a television cabi-
net and went into an irreversible coma. As she was already dying from inop-
erable meningioma brain tumors—after enduring a decade of craniometry
brain surgeries, Gamma Knife radiation treatments, and chemotherapy—in
the course of a few weeks my father and I decided to pull her feeding tube
and start death's clock, which wound down ten days later. As we were with
her most of that time, I could see and sense her discomfort. Ten days without any nutrition must surely have been an unpleasant experience. Her
mouth was so dry that I periodically gave her water to slake her thirst. Her
nurse gave her sponge baths so that she would feel clean. Her docs had her
on painkillers just in case there was any undetected internal distress—who
knows what goes on inside a comatose mind? She was supposedly unconscious, but I could communicate with her by holding her hand and asking
her questions, to which I would occasionally receive a squeeze in response.
She was in some state of altered consciousness, so I repeated over and over,
"I love you, Mama. You are a great mother and a heroic person. Wise owl,
brave soul. A life well lived. It's okay to say good-bye."

Someone was still in there, and given her obvious discomfort—
however fleeting it might have been in her state of mind—I wondered if
it might not have been more humane to bring about my mom's death
sooner. But if so, how? Lethal injection or some other Kevorkianesque
technique? That would be illegal. What about smothering her with a pil-
low? That would be swift and less likely to leave legally convicting evidence. Yikes! The very thought is repulsive. I could barely even write
those words. Why is there an intuitively visceral difference between allowing someone to die and bringing about her death, or between passive
and active euthanasia?[2]

These feelings we have about right and wrong, moral and immoral, good and evil, are telling us something important about how we should behave, even if the moral calculation seems irrational on the surface. In my previous book, *The Science of Good and Evil*, I argued that we have an *evolved moral sense*, by which I mean moral feelings or moral emotions. For example, positive emotions such as righteousness and pride are experienced as the psychological feeling of doing "good." These moral emotions likely evolved out of behaviors that were reinforced as being good either for the individual or for the group. Negative emotions such as guilt and shame are experienced as the psychological feeling of doing "bad." These moral emotions probably evolved out of behaviors that were reinforced as being bad either for the individual or for the group. The moral sentiments represent something deeper than specific feelings about specific behaviors. While cultures may differ on what behaviors are defined as good or bad, the general moral emotion of feeling good or feeling bad about behavior X is an evolved emotion that is universal to everyone.[3]

What I am after here is not why someone would feel guilty about some specific violation of a social norm—such as lying or stealing—but why anyone should feel guilty about anything. The feeling of guilt must have some deeper purpose that goes beyond immediate cultural norm violations. Emotions require brain power, and brains are expensive to run, so for evolution to have created a powerful emotion, there must have been a reason for it. Two of our more basic emotions serve as examples: hunger and sexual arousal.

When we need energy we do not consciously compute caloric input/output ratios; we simply feel hungry, and that emotion triggers eating behavior. When we need to procreate we do not consciously calculate the genetically based health indicators of potential sexual partners; we simply feel attracted toward someone, and that emotion triggers sexual arousal and behavior. For example, we are attracted to people whose bodies and faces are bilaterally symmetrical—where the left and right sides match each other fairly closely. Women are attracted to men with an inverted-pyramid-shaped upper body (narrow waist and broad shoulders), and men are attracted to women with a waist-to-hip ratio of 0.7:1. It turns out that these characteristics—along with full lips, strong cheek bones, thick and silky hair, and an overall hourglass shape in women, height, hair, and a strong jaw in men, and clear complexion in both men and women—are indicators of good genetic health.[4] Now, no one walks into a room and starts computing bilateral symmetry and waist-to-hip ratios. Evolution has done the computation for us, and in its stead produced emotions that are proxies for these computations. In other words, we are hungry and aroused because, ultimately, the survival of the species depends on food and sex, and those organisms for which healthful foods tasted delicious and for which sex was exquisitely delightful, left behind more offspring. What we inherit are emotions that guide our behaviors.[5]

The economics of sexual emotions also has a deeper evolutionary basis, as discovered in such cross-cultural studies as the International Mate Selection Project, which examined thirty-seven cultures from six continents and five islands, in which people were asked to rate members of the opposite sex on eighteen different characteristics of mate desirability. Men from all over the world preferred younger attractive women of prime reproductive age with the physical characteristics noted above, whereas in 36 of the 37 samples, women valued men's "good financial prospects" over "good looks." Further, the study found that women tend to marry older men who have more resources, more attractive women were more likely to be married to men of higher occupational status, and that those who follow this pattern tend to have more children than those who follow a different mating strategy.[6]

An additional source of evidence for the evolution of the moral emotions comes from the cross-cultural study of human universals, or those features of human thought, behavior, language, social relations, and culture, for which there are no known exceptions in any human societies past or present.[7] The most common examples of universals include tools, myths and legends, sex roles, social groups, aggression, gestures, grammar, and emotions. The anthropologist Donald Brown has compiled a comprehensive list of 373 human universals, from which I count 202 (54 percent) directly related to morality and religion. From this list it is strikingly clear just how much of what we do has some bearing on our state of being as social organisms in interaction with others of our kind.

Some universal moral emotions include *affection expressed and felt* (necessary for altruism and cooperation), *attachment* (necessary for bonding, friendship, mutual aid), *coyness display* (courtship, moral manipulation), *crying* (expression of grief, moral pain), *empathy* (necessary for moral sense), *envy* (moral trait), *fears* (basis of guilt), *generosity admired* (reward for cooperative and altruistic behavior), *incest taboo* (moral prohibition with genetic implications), *judging others* (foundation of moral approval/disapproval), *mourning* (expression of grief), *pride* (a moral sense), *self-control* (moral behavior), *sexual jealousy* (foundation of moral mate guarding), and *shame* (moral sense).

Some universal moral behaviors include *age statuses* (social hierarchy, dominance, respect for elder wisdom), *coalitions* (foundation of social and group morality), *collective identities* (basis of xenophobia, group selection), *conflict mediation* (foundation of much of moral behavior), *customary greetings* (part of conflict prevention and resolution), *dominance/submission* (foundation of social hierarchy), *etiquette* (enhances social relations), *family (or household)* (the most basic social and moral unit), *food sharing* (form of cooperation and altruism), *gift giving* (reward for cooperative and altruistic behavior), *government* (social morality), *group living* (social morality), *groups that are not based on family* (necessary for higher moral reasoning and indirect reciprocity), *inheritance rules* (reduces conflict within families and communities), *institutions* (rule enforcement), *kin groups* (foundation of kin selection/altruism and basic social group), *law (rights and obligations)* (foundation of social harmony), *marriage* (moral rules of foundational relationship), *reciprocal exchanges* (reciprocal altruism), *redress of wrongs* (moral conflict resolution), *sanctions* (social moral control), *sanctions for crimes against the collectivity* (social moral control), and *sanctions that include removal from the social unit* (social moral control).

Finally, some universal economic emotions and behaviors, based on the fundamental principle of reciprocity universally expressed as the golden rule of "do unto others as you would have them do unto you," include *cooperative labor* (part of kin, reciprocal, and indirect altruism), *fairness* (equity), *food sharing* (form of cooperation and altruism), *generosity admired* (reward for cooperative and altruistic behavior), *gestures* (signs of recognition of others, conciliatory behavior), *gift giving* (reward for cooperative and altruistic behavior), *hospitality* (enhances social relations), *insulting* (communication of moral disapproval), *judging others* (foundation of moral approval/disapproval), *planning for future* (foundation for moral judgment), *pride* (a moral sense), *promise* (moral relations), *negative reciprocity* (revenge, retaliation; reduces reciprocal altruism), *positive reciprocity* (enhances reciprocal altruism), *redress of wrongs* (moral conflict resolution), *shame* (moral sense), and *turn-taking* (conflict prevention).

What these universals reveal is that we are social primates, moral primates, and economic primates, and thus these characteristics belong to the species and transcend the individual members of our species.

◆ ◆ ◆

From such basic emotions as hunger, arousal, and mate choice to the higher social practice of monogamy and the economic institution of marriage, we cross the bridge from evolutionary psychology to evolutionary economics. We evolved as a pair-bonded primate species. Although anthropologists have classified nearly 80 percent of human societies as practicing polygyny (more than one wife), since the sex ratio is nearly fifty-fifty, in actual practice only a few men have multiple wives, while the vast majority of people are coupled monogamously.[8]

Yet within this arrangement we see a dramatic difference in the reproductive strategies followed by men and women. As the Billy Crystal character in the film *City Slickers* quipped, "Women need a reason to have sex; men just need a place." As I described in my book, *Why Darwin Matters*, this observation was borne out in an amusing study by the psychologists Russell Clark and Elaine Hatfield, who had an attractive member of the opposite sex approach a fellow college student whom he or she had not previously met and ask one of three questions:

1. Would you go out on a date with me tonight?
2. Would you go back to my apartment with me tonight?
3. Would you sleep with me tonight?

The results will elicit a wry smile even while surprising no one: To the first question, half of both men and women responded positively; to the second question, however, 69 percent of one gender and only 6 percent of the other gender agreed to return to the apartment; and to the third question, 75 percent of one gender agreed to have sex, while 0 percent—not one—of the other gender agreed to a proverbial "zipless fuck," which even

Ms. Jong admitted was rarer than a unicorn.[9] I don't even need to say which gender was which. You already know.

Think of this evolutionary analysis in economic terms. Life, like the economy, is about *the efficient allocation of limited resources that have alternative uses.* Sperm, which are tiny and plentiful, are virtually unlimited, whereas eggs, which are large and scarce, are exceedingly limited; thus, women far more than men need to be concerned about allocation efficiency. The result is that men compete among themselves for access to women, while women do the selective choosing. Darwin called this *sexual selection,* and it is a powerful force in evolution.

My point is this: as we saw with the emotions of hunger and arousal, you do not need to compute the economic value of choosing a marital partner; let your emotions guide you. An emotion such as love evolved to tell you if the person you are considering for marriage is healthy, trustworthy, faithful, reliable, and stable; that is, will he or she make a good father or mother for your children? Conversely, a negative emotion such as jealousy is a proxy for evolution's calculation that it is highly inefficient to invest in someone who is allocating scarce resources to an alternative partner—also known as cuckolding.

Consider how monogamy, adultery, and jealousy play out in an evolutionary analysis. If monogamy had no benefits at all, then we would not practice it. But we do. And there are deep evolutionary reasons why. For example:

- Women practice monogamy as the best reproductive strategy because eggs are limited and bringing a fetus to term and raising a child to reproductive age is a huge investment to make, so women tend to be very selective about choosing a mate and deeply invested in making the pair bond last.[10]
- Most men practice monogamy because the sex ratio is close to fifty-fifty and even in polygamous societies only a few men have multiple wives.
- Although polygamy and promiscuity afford men greater *quantities* of reproductive opportunity (in other words, more women), some evolutionary biologists show how the most profitable mating strategy for men is to ensure the *quality* of a few offspring by investing

heavily in one mate, rather than taking the risk of raising some other man's offspring by being promiscuous and investing in many children.[11]

• Monogamy protects against sexually transmitted diseases (STDs) that were almost certainly a part of our Paleolithic past.

• Stepchildren suffer significantly higher rates of physical and psychological abuse at the hand of stepparents, especially stepfathers, than do children from biologically intact families.[12]

• Stepchildren leave home significantly earlier than children in intact families.[13]

• Stepchildren show higher rates of behavioral, emotional, and physical problems compared to children in biologically intact families.[14]

• Stepdaughters in particular are at a significantly greater risk of being sexually molested by their stepfathers than daughters are by their biological dads.[15] The incest taboo, so powerful in squelching feelings of sexual attraction between relations, becomes a less potent prophylactic against stepfathers, who may have been introduced to the girl well after the incest taboo imprinting period has passed.

• Young single males are more prone to engage in risky, competitive, and even violent behavior as a means of seeking status, controlling females, and competing with other males. Cross-cultural data show that crime rates are higher in polygamous than monogamous societies,[16] and that men are twenty times more likely to be killed by another man than a woman is of being killed by another woman.[17] Even in modern America, a 1985 study found that 41 percent of adult male offenders were unemployed and 73 percent were unmarried, and that men in general committed 93 percent of robberies, 94 percent of burglaries, and 91 percent of car thefts.[18] As the evolutionary psychologist Steven Pinker likes to say, the number one predictor of crime and violence is *maleness!*[19] But unmarried maleness makes it even worse. These guys are competing for status and resources in order to attract females. Once they've attained marital stability, such behavioral expressions are attenuated.

In like manner, if adultery had no benefits at all, then we would not practice it. But we do. And there are deep evolutionary reasons why.[20] For

men, adultery provides an opportunity to allocate one's virtually unlimited genes with alternative partners. For women, adultery is a chance to trade up for superior genes, better resources, and higher social status. How do we square the circle of monogamy and adultery? Serial monogamy. Most people, most of the time, in most circumstances practice monogamy. Occasionally some people, some of the time, in some circumstances practice adultery. The balance is heavily weighted in favor of monogamy, but not exclusively so, and the reason has to do with the relative risks for both monogamy and adultery. For monogamy, there is the risk that you made a bad choice: your partner has unhealthy genes, is infertile, is untrustworthy, unfaithful, unreliable, or unstable, and thus will not make a good parent to your children. For adultery, the hazards are even more serious. For men, revenge by the adulterous woman's husband can be deadly, and although he is not likely to be killed by his wife if she catches him cheating on her, she can impose significant emotional and social penalties, such as loss of contact with children, depletion of financial resources, social shunning, and risk of sexual retaliation that thereby increases the odds that if he stays he may be allocating his resources toward another man's offspring. For women, being discovered by the adulterous man's wife involves little physical risk, but getting caught by one's own husband can and often does lead to extreme physical abuse and occasionally even death—research shows that most spousal murders are triggered by sexual jealousy.[21]

Finally, for children, being raised by nonbiological parents turns out to be one of the riskiest and even deadliest factors ever discovered in research on childhood neglect and abuse—Martin Daly and Margo Wilson found that a child living with one or more stepparents was 100 times more likely to die from abuse than a child raised by biological parents.[22] (As in the saying, "beaten like a red-headed stepchild.") Perhaps this is why a study on the response of extended family members commenting on the resemblance of newborn babies to their parents found that the family members—most notably those on the mother's side—were likely to voice their opinion that the newborn looks like the father.[23] This makes sense in the context of additional studies on genetic paternity that found that anywhere from 5 to 30 percent of husbands in a maternity ward were not the biological fathers of the infants they were holding.[24] No wonder Shakespeare waxed poetic as he did in *Othello*:

O, beware, my lord, of jealousy;
It is the green-ey'd monster which doth mock
The meat it feeds on.

That green-ey'd monster was born out of an evolutionary context—jealousy is an evolved emotion that stands in for an evolutionary computation that has weighed the relative costs and benefits of sexual infidelity. This has produced the emotional tension most of us have felt between the desire for sexual variety and the fear of the consequences of getting caught acting on that longing. Here again we see the role of such evolved emotions as guilt or pride at having done the wrong or right thing in interplay with these other evolved emotions of love and jealousy. And it is here that morality comes into play, as we make the choice to act on our emotions or hold them in check. I have made the case elsewhere that we have free will and thus can genuinely make moral choices;[25] and since this choice has economic consequences, it is an example of virtue economics at work, the virtue here being fidelity, in the economic institution of marriage.

✦ ✦ ✦

On many mornings, I go on training ride with a group of competitive cyclists in the Southern California area. My routine is to cycle from my office to the start of the ride in order to get in some extra mileage, often wending my way through the congested streets of Glendale to get to the open roads of Griffith Park. One morning, a black Town Car parked on my side of the road caught my eye because the driver was just sitting in the car while what appeared to be his passenger—an elderly and frail-looking woman—was slumped on the sidewalk, struggling to get up. Without even a whisper of conscious thought, I hit the brakes, hopped the curb, dismounted my bike, reached down and slipped my arms beneath the woman's arms, and hoisted her up to her feet. Once she was stable and able to walk, she thanked me and went on her way, as did I.

A sizable body of psychological literature is devoted to understanding why people are violent, aggressive, malicious, and mean. A far smaller body of literature is devoted to explaining why people are nonviolent, gentle, kind, and compassionate. For every act of violence that reaches the

highlight reels of the nightly news, ten thousand acts of kindness go unrecorded and, in most cases, unnoticed in the background noise of general human virtue. Like the fish that does not even notice the water in
which it resides, the virtues of our humanity are such a common occurrence that we swim blissfully unaware in its invisible waters. From a
"selfish gene" perspective, competitiveness and greed need no particular
explanation beyond the obvious—I'll do whatever I can to get my genes
into the next generation, even if that means stomping on all the little people on my way to the top. But if helping others by being selfless and altruistic decreases the chances of getting my genes into the next generation,
why would I do it? The short answer is that it is a myth that evolution is
driven by selfishness; it is, in fact, driven by *adaptability*, and in a social
primate species like ours, more often than not the most adaptable thing
you can do to survive and reproduce is to be cooperative and altruistic.

✦   ✦   ✦

Over the course of millions of years, our moral emotions have evolved,
primarily under biological control. In the early stages of our evolution,
the individual, family, extended family, and small groups were molded
primarily by natural selection, and the individual's need for survival and
reproduction is met through the family, extended family, and local group
members.

But around thirty-five thousand years ago, a transition took place, and
cultural factors increasingly assumed control in shaping our moral behavior. In later stages, communities and societies were shaped primarily
by cultural selection. Basic psychological and social needs—security,
bonding, socialization, affiliation, acceptance, affection—evolved to aid
and reinforce cooperation and altruism, and all facilitate genetic propa-
gation through children. Such *kin altruism* works indirectly—siblings
and half siblings, grandchildren and great-grandchildren, cousins and
second cousins, nieces and nephews, all carry portions of our genes.[26]
Anyone who is genetically related to us is included. In larger communities and societies, in which an individual has no genetic relationship to
most others, *reciprocal altruism* (I'll scratch your back if you'll scratch
mine) and *blind altruism* (if you scratch my back now, I'll scratch yours
later) are needed to supplement kin altruism. We become less inclusive

and more exclusive.[27] The more a moral emotion reaches beyond ourselves, the further it goes in the direction of helping someone genetically less related, the less support it receives from underlying evolutionary mechanisms.

In this social context, fairness evolved as an Evolutionary Stable Strategy (ESS) for maintaining social harmony in our ancestors' small bands, where cooperation was reinforced and became the rule and freeloading was punished and became the exception. For example, in experimental economics research utilizing exchange games in which cooperation and defection are both strategies in which subjects can gain or lose depending on what the other subject does, the most successful—and thus most commonly employed—strategy is called "tit for tat," where you start off cooperating and continue cooperating as long as your partner does, but punish defectors. The theory holds that an ESS will be selected for because it leads to greater survival by its practitioners, and will thus be passed along to future generations.

The reason hunter-gatherer bands are egalitarian is not because they are naturally altruistic or lack some impulse for competitiveness and avarice, but because excessive greed and selfishness is kept in check by the fellow group members. Anyone who attempts to hoard food, amass tools and other products, or steal someone else's spouse is likely to be shunned or punished by the rest of the group, and unless that individual is a high-status "big man" with an inner circle of loyal followers who must, in turn, be well nourished and paid for their services, excessive selfishness is kept in check by the collective power of the group. Being ousted from a tiny band of hunter-gatherers in the Paleolithic was probably a death sentence, unless one could ingratiate with another group. And if you have a reputation for being self-centered and socially unreliable, you will need to travel far to find people who do not know your reputation.

Evidence for this social enforcement of egalitarianism can be seen in a number of anthropological studies of meat-sharing practices in modern hunter-gatherer societies around the world. It turns out that these small communities—which can cautiously be used as a model for our own Paleolithic ancestors—are remarkably egalitarian. Using portable scales to measure precisely how much meat each family within the group received after a successful hunt, researchers found that the immediate families of successful hunters got no more meat than the rest of the families in the group, even when these results were averaged over several weeks of regular hunting excursions. Hunter-gatherers are egalitarian because selfish acts are effectively counterbalanced by the combined will of the rest of the group. The anthropologist Chris Boehm has discovered in a number of hunter-gatherer societies the use of gossip to ridicule, shun, and even ostracize individuals whose competitive drives and selfish motives interfere with the overall needs of the group.[28] In other words, we are competitive and selfish, but we are also cooperative and altruistic—tendencies that are created and reinforced by the group in which the individual lives. In this manner does a human group become a moral group in which "right" and "wrong" coincide with group welfare and self-serving acts, respectively.[29]

Some of this cooperative behavior may be accounted for by reciprocal altruism or *inclusive fitness*, but a deeper interpretation is that an emotional sense of "right" and "wrong" action evolved in humans living in hunter-gatherer communities through genetic transmission of such traits, as well as their cultural spread through modeling and learning. An anthropological example of how this process works can be seen in the Malaysian rain forest tribe called the Chewong. Like other hunter-gatherer groups, the Chewong (who also employ limited agriculture) are egalitarian, a way of life that is governed by a system of superstitions called *punen*. In the words of the anthropologist Signe Howell, who has studied the Chewong, *punen* can be defined as "a calamity or misfortune, owing to not having satisfied an urgent desire."[30] In the Chewong world, strong desires are connected with food, and the powerful norms about food sharing are associated with the myth about "Yinlugen bud," who supposedly brought the Chewong out of a more primitive state by insisting that eating alone was improper human behavior. Myth, gods, religion, and morality are all integrated in the Chewong culture by the concept of *punen*, and are linked to a most practical matter of individual and group survival—eating and sharing of food. Thus, says Howell, the Chewong avoid provoking *punen* at all costs. When food is caught away from the village, it is promptly brought back, publicly displayed, and equitably distributed among all households and among all individuals within each home. To reinforce the sanction against *punen*, someone from the hunter's family touches the catch, then proceeds to touch everyone present, repeating the word *punen*. In this system religious superstitions and gods oversee the exchange process, generating within the individuals an overall sense of right and wrong action as related to the success or failure of the group.

My point with this example is that we do not need to make social and moral computations about what is right and wrong. Evolution has done the calculating for us, and our emotions guide our behaviors. This is why being shunned by one's family, extended family, or social circle feels so bad, and it shows the power of emotions to guide our social choices, such as deciding to be fair in an economic exchange.

✦ ✦ ✦

Kin selection, inclusive fitness, and reciprocal altruism explain most of the behaviors we consider to be moral toward others—relations and strangers. But there are some actions that have no apparent benefit to the giver and cannot be reciprocated by the receiver, such as handouts to beggars, donations to nonprofit charities, donating blood, "adopting" a child in another country whom you will never meet, and the like. The Israeli evolutionary biologists Amotz and Avishag Zahavi suggest a mechanism they call *Costly Signaling Theory* (CST) to explain such kinds of altruistic acts.[31] In a CST model, people sometimes do things not just to help those related to them genetically, and not just to help those who will return the favor, either now or later, but to send a signal, or a message, that says, in essence, "My altruistic and charitable acts prove that I am an honest and trustworthy member of the community, and that I am so successful that I can afford to make such sacrifices for others and for the group." That is, altruism is a form of information that carries a signal to others of trust and status—*trust* that I can be counted on to help others when they need it so that I can expect others to do the same for me; *status* that I have the health, intelligence, and resources to afford to be so kind and generous. To repeat my earlier point, it is not enough to fake being a moral person (because others will find you out eventually); you have to be (or believe you are) a moral person.

In yet another line of evidence that we evolved moral emotions, brain

scans reveal that we evolved powerful neurological mechanisms to reinforce cooperation and prosocial behavior, and through social exchange
create and reinforce bonds between unrelated people.[32] The neuroeconomist Kevin McCabe, for example, scanned the brains of subjects participating in a "trust and reciprocity" game, revealing that areas of the
prefrontal cortex related to impulse control and the delay of immediate
gratification are more active in the brains of cooperators than defectors,
suggesting that cooperation requires "attention to mutual gains with the
inhibition of immediate reward gratification to allow cooperative decisions."[33] In her study of both humans and the great apes, neuroscientist
Katerina Semendeferi found that Area 10 of the frontal lobe is associated with such higher cognitive functions as the undertaking of initia-
tives and the planning of future actions, and that this area, while larger
in apes than in monkeys, is in humans the largest of all the apes and is
more connected to higher order brain areas. She concludes that "the
neural substrates supporting cognitive functions associated with this
part of the cortex enlarged and became specialized during hominid evolution."[34]

The reason for this cortical expansion is that humans evolved to become the preeminent social and moral primate. Another neuroscientist,
Jorge Moll of the National Institutes of Health, and his colleagues, for
example, found that moral emotions activate both the *amygdala* (emotions) and the orbital and medial prefrontal cortex (cognitions), showing
that moral acts are as much a function of moral emotions as they are of
moral reasoning.[35] In a subsequent study on charitable donations, Moll
found that when subjects had an opportunity to donate to or oppose real
charitable organizations related to prominent social causes, the mesolimbic reward system was activated by donations in the same manner as
when subjects are rewarded with monetary gains for a task. In other
words, the moral feeling that comes from being charitable is directly analogous to the feeling of getting paid for any other act.[36]

Since morality principally involves our responses to others in social
situations, we cannot separate the moral from the social. In order to be a
moral agent one must be both self-aware and aware that others are self-
aware, functions that are located in two different areas of the brain. Self-
awareness, at least in part, appears to be located in the medial prefrontal

cortex, whereas representing others' actions and intentions appears to be centered in the temporal cortex.[37] There even appear to be specific neurons in the brain evolved for social and moral emotions.

✦ ✦ ✦

Since we are a social primate species, one would expect that other social primate species have evolved specialized cells for dealing with others in social situations. One candidate is a special class of cells called *spindle cells*, or *Von Economo neurons*, first discovered in 1925 by the Romanian-Austrian brain anatomist Constantin von Economo. Cigar-shaped and tapered at each end, spindle cells are located in just two regions of the frontal lobes: (1) the *anterior cingulate cortex* (ACC), an evolutionarily ancient region of the brain common to all mammals that lies beneath the midline of the cerebral cortex, and (2) the *frontoinsular cortex* (FIC), just behind the eyes. Spindle cells are also the exclusive property of the great apes: humans, chimpanzees, bonobos, gorillas, and orangutans. They have not been found in monkeys or other mammals. According to the Caltech primatologist and brain specialist John Allman, spindle cells appear to be unique to our evolutionary line, and the more distant the ape relation, the lower the spindle cell count. Orangutans have the least, humans the most, with chimps, bonobos, and gorillas in between.

The significance of spindle cells for social emotions is inferred from their location. Brain scan studies have shown that the ACC guides attention, senses pain and errors, taps into the body's autonomic control systems, and generally acts as a central hub between thoughts and emotions. The FIC is especially active in response to others in social situations, such as in a mother when she hears her infant cry, when we observe others whom we love experiencing pain, and even when we are being deceived. "All these responses have something in common," Allman explained. "They all represent value judgments within a social context. I think spindle cells are the home of the complex social emotions." The size of the spindle cells is key, especially in their large axons. Neural axons carry signals to other neurons, and the larger the axons, the faster the signal. And speed is the key to rapid cognition in social situations.

In other words, our emotions are guiding our decisions, and doing so in the blink of an emotional moment—"this feels right" or "this feels wrong." This fits my model of moral emotions. As Allman noted, "The main thing spindle cells do is to adjust your behavior in a rapid real-time interaction in a complex social environment. It is so simple that I think it is entirely reasonable that it could be performed by about 100,000 neurons." And spindle cells feature receptors for the neurotransmitters serotonin, dopamine, and vasopressin, associated with reward value, bonding, and love. Thus, Allman suggests that spindle cells evolved sometime in the past ten million years, presumably before the ape-human split some six to seven million years ago.[38]

The final step in establishing a deep evolutionary basis for our moral emotions is to identify the underlying neural architecture that generates behaviors and feelings involved in acting moral. We now have that foundation in the form of *mirror neurons*, specialized neurons that "mirror" the actions of others and thus are directly involved in imitation, anticipation, and empathy, all key ingredients in the evolution of the moral emotions.

In the late 1980s and early 1990s, the Italian neuroscientist Giacomo Rizzolatti and his colleagues at the University of Parma discovered mirror neurons serendipitously when they were recording the activity of single neurons in the *ventral premotor cortex* of macaque monkeys. Poking hair-thin electrodes into individual neurons allows neuroscientists to monitor the rate and pattern of single cell activity, and in this case the action from the monkey's F5 neurons spiked whenever it reached for a peanut placed in front of it in the experimental apparatus. The serendipity came when one of the experimenters reached in and grabbed one of the peanuts, and the same neurons in the monkey's brain fired. Monkey do was the same as monkey see. The motor neurons were "mirroring" the motor activity of others—hence they became known as mirror neurons. As Rizzollati recalled, "We were lucky, because there was no way to know such neurons existed. But we were in the right area to find them."[39]

Neuroscientists throughout the 1990s scrambled to learn more about mirror neurons, finding them in other parts of the brain, such as the *inferior frontal* and *inferior parietal* regions of the brain, and not only in monkeys but in humans as well.[40] Since research ethics committees generally frown upon open brain research on human subjects, individual mirror neurons have not been monitored in people, but employing the next best technology—fMRI—UCLA neuroscientist Marco Iacoboni and his colleagues imaged the brains of subjects as they watched people make finger movements and then imitated those same finger movements, discovering that the same areas of the frontal cortex and parietal lobe in both conditions were active.[41]

The purpose of mirror neurons, and why they would have evolved, is a subject of controversy. Rizzolatti's original thesis—that mirror neurons are just motor neurons responding to seeing as well as doing—is uncontroversial and makes sense. When you see an action, it is recorded on your visual cortex, but to more deeply understand what the act means in terms of its consequences, the observation must be linked to the motor system of the brain so that there is an internal check with the external world. With this basic neural network in place, higher order functions can be layered onto it, such as imitation. In order to imitate someone's actions, you need both a visual memory of how the action looks and a motor memory of how the action feels when implemented, and there is now considerable research linking the mirror neural network to imitation learning.

Being able to imitate others, however, can go much deeper than the mere repetition of their actions.[42] Actions have intentions, and so mirror neurons have also been implicated in what is called *Theory of Mind* (ToM), or the ability to understand that others have beliefs, desires, and intentions. A higher order ToM allows one to realize that others' intentions may be the same as or different from your own. This is sometimes called "mind reading," or the process of inferring the intentions of others by projecting yourself into their minds and imagining how you would feel. A still higher level ToM means that you understand that others also have a theory of mind, and that you know that they know that you know they have a theory of mind. Here we may find ourselves in an intentional loop not unlike a bit from the 1950s television series *The Honeymooners*, in which Jackie Gleason tells Art Carney, "You know that I know that you know that I know that . . ." But how does ToM mind reading actually operate in the brain?

In a review of the research on what brain scans have revealed about the location of Theory of Mind, Glasgow University neuroscientists Helen Gallagher and Christopher Frith conclude that there are three areas consistently activated whenever ToM is needed: the *anterior paracingu-*

late cortex, the *superior temporal sulci*, and the *temporal poles* bilaterally. The first two brain structures are involved in processing explicit behavioral information, such as the perception of intentional behavior on the part of other organisms—for example, "That wolf intends to eat me." The temporal poles are essential for the retrieval from memory of personal experiences, such as "The last time I saw a wolf, it tried to eat me." All three of these structures are necessary for ToM, and Gallagher and Frith go so far as to posit that the anterior paracingulate cortex (located just behind your forehead) is the seat of the Theory of Mind mechanism.[43] Theory of Mind is a high-road automatic system that kicks in for specified activities involving other people, particularly in social situations. It most likely evolved out of a number of preexisting neural networks used for other related activities, such as the ability to distinguish between animate and inanimate objects, to hold the attention of another through eye gaze, the ability to distinguish the actions of self and others, and the ability to represent actions that are goal directed. All of these functions are basic to survival in any social mammal, and thus Theory of Mind is most likely an exaptation, a feature co-opted for a different purpose from the one for which it was originally evolved.

A number of specific brain scan studies support this idea. In one 1998 fMRI experiment, people were shown two different hand actions, one without a context and one with a context that revealed the intention of the action. The latter scene activated the subject's mirror neuron network, revealing where ToM is located in the brain.[44] In 2005, a very clever experiment was conducted in which monkeys watched a person either grasp an object and place it in a cup or grasp an apple and bring it to his mouth—similar action, different intention. Recording forty-one individual mirror neurons in the inferior parietal lobe of the monkeys' brains, it was discovered that the "grasp-to-eat" motion triggered fifteen mirror neurons to fire, but these were silent when observing the "grasp-to-place" motion. Yet four other mirror neurons were active for the opposite condition. Interestingly, the neuroscientists concluded, the mirror neurons in this part of the brain "code the same act (grasping) in a different way according to the final goal of the action in which the act is embedded."[45] In other words, there are neurons specialized for discriminating between different intentions: grasping in order to place versus grasping in order to eat. More generally, this implicates mirror neurons in both predicting others' actions and inferring their intention.

From imitation to intention to emotion, additional fMRI studies point to a mirror neuron network involved in empathy. In one such study, Christian Keysers and his neuroscientist colleague Bruno Wicker scanned the brains of fourteen subjects as they were exposed to two different conditions: (1) a disgusting odor (butyric acid, which gives off the smell of rotten butter) and (2) short clips of people making facial expressions of disgust. Wicker and Keysers found that the feeling of disgust and watching someone else express disgust activated the same network of neurons called the anterior insula, involved in the processing of emotion. It appears that experiencing disgust and seeing disgust are indistinguishable to the insula. In a related study, they examined "tactile empathy" and found that lightly touching someone on the leg activated the same area of the *somatosensory cortex* as when they were shown photographs of someone being touched in the same spot.[46] It appears that experiencing touch and seeing touch are indistinguishable to the somatosensory cortex. A related fMRI experiment confirmed the finding for more general emotions, linking the observation and imitation of facial expressions in the anterior insula—seeing someone's facial expression registered the same activity in the insula as making the expression yourself.[47]

To further refine the difference between the imitation of motion and the empathy of emotion, Marco Iacoboni and his colleagues scanned the brains of twenty-three subjects as they watched videos of a hand picking up a teacup. In the first clip, a hand is reaching for the cup sitting on a well-set table, implying that the intention was to grasp the cup and take a sip. In the second clip, the table is messed up, with cookie crumbs strewn about, implying that the intention was to clear the table. As a control, a third clip shows the hand reaching for the cup by itself, with no context. The results revealed a distinct difference between the first two clips and the third clip, with strong activity in the mirror neuron network of the *premotor cortex* for the intention versus no-intention scenes, and a stronger signal for the drinking condition than the cleaning condition. As Iacoboni concluded: "The stronger activation of the inferior frontal cortex in the 'drinking' as compared to the 'cleaning' intention condition is consistent with our interpretation that a specific chain of neurons coding a probable sequence of motor acts underlies the coding of intention."[48]

As with all other human traits, empathy varies among individuals, and the research on mirror neurons lends neural credence to this folk observation. People who score high on self-report questionnaires measuring empathy also show stronger activity in mirror neuron networks for both movement and emotions. Likewise, women show stronger EEG readings linked to mirror neuron activity than men do, lending scientific credibility to the intuition that women are more empathetic than men.[49] As for a more literal meaning of empathy, research on observed versus felt pain, particularly when the pain was imposed on someone who is loved by the observer, activated the same brain circuitry.[50] Empathy has its very own brain network. Adam Smith would not have been surprised.

There is also some evidence that a dysfunctional mirror neuron network might be involved in autism, because of the difficulty autistic children have in relating to people and social situations. When shown a short clip of a hand making grasping motions and then asked to imitate that grasping motion, autistic children show dramatically different EEG readings than nonautistic children. If you can't read others' intentions, then their actions will appear random and thus meaningless, requiring no response, or a random and thus inappropriate response.[51] Sound similar to trying to read others' intentions in the stock market?

The research on mirror neurons and autistic children was conducted by the always colorful and cleverly creative University of California, San Diego, neuroscientist V. S. Ramachandran. "Rama" (to those who know him) has taken the implications of mirror neuron research and run with it to the high table of evolutionary theory. Calling the discovery of mirror neurons "the single most important 'unreported' (or at least, unpublicized) story of the decade," Rama holds that mirror neurons are one of the key steps in making humans different from all other animals, in that by being able to imitate, we can override natural selection—and consequently we can modify the environment instead of the environment modifying us. In a period of global cooling, for example, natural selection will favor those animals with more effective thermoregulation features, such as a thick fur coat. But this can take hundreds or thousands of generations to evolve, whereas a big-brained primate with a mirror neuron network can observe a thick-furred mammal looking toasty warm in a blizzard and immediately grasp the idea of killing the animal and wearing its fur coat rather than waiting around for evolution to grow him one. Others see and do what our erstwhile hominid ancestors discovered, and with language we can quickly spread the fur-coat meme (along with the stone-tool meme, the fire meme, the bow-and-arrow meme, and the art, music, and religion memes) throughout Cro-Magnondom. It's a bit of a just-so story, but one that has merit given what we know about the power of imitation, intention, and empathy in our neural architecture.[52]

From the neurophysiology of empathy, we can begin to understand another biological basis of economics. "Understanding the intentions of others while watching their action is a fundamental building block of social behavior," Marco Iacoboni reflected. "Our findings show for the first time that intentions behind actions of others can be recognized by the motor system using a mirror mechanism in the brain. The same area of the brain responsible for understanding behavior can predict behavior as well." In other words, there is empirical evidence for Adam Smith's philosophical deductions in *The Theory of Moral Sentiments*: "How selfish soever man may be supposed, there are evidently some principles in his nature, which interest him in the fortune of others, and render their happiness necessary to him, though he derives nothing from it except the pleasure of seeing it. Of this kind is pity or compassion, the emotion which we feel for the misery of others, when we either see it, or are made to conceive it in a very lively manner."[53]

◆ ◆ ◆

Addressing such problems as preserving the planet's ecosystem and biodiversity or maximizing within-group amity and minimizing between-group enmity requires social and political action. Unfortunately, such high-minded goals are too far-reaching, and the time frames involved are too long-range, for how we were programmed by nature to think. In the Paleolithic environment, our concern for the environment and biodiversity was restricted to a few tens of kilometers and hundreds of species over a lifetime of only a few decades. The number of people our ancestors encountered in their lifetimes could be numbered in the hundreds,

so there was no reason for evolution to produce a principle of tolerance for ethnically diverse members of other groups in faraway lands.

This is why scientists and social activists concerned about global climate change and other long-term environmental threats have had such a difficult time getting people to notice the problem, let alone care about finding a solution. When your evolved moral sense of what is important is restricted to a few dozen people, miles, and years, who cares what happens to some other group thousands of miles away or a thousand years hence?

One answer is to reframe the problem in an intimate, short-term context. Consider how nonprofits concerned with the plight of starving third-world children employ the "adopt a child" strategy in order to tap into potential first-world donors' natural empathies. The effect of the strategy was endearingly portrayed in the 2002 film *About Schmidt*, in which Jack Nicholson's title character adopts a Tanzanian child named Ndugu, with whom he carries on a one-way correspondence that becomes the narrative outline of the story of Schmidt's search for meaning in his later years. After writing countless self-centric letters about matters trivially irrelevant, Schmidt discovers in the film's final scene that his foster charge cannot read or write. But the letter from the nun who looks after the boy brings redemption, as it is accompanied by a stick-figure drawing made for Schmidt by Ndugu that depicts an adult and child holding hands beneath a deep blue sky and radiant yellow sun. The scene is so moving in its emotional simplicity that it evokes empathetic tears.[54] By touching one small child worlds away—a child with a name and a face and a visual acknowledgment for a small but significant act of kindness—Schmidt's life became meaningful. Call it the Ndugu Effect.

We care more about one named child with a face than we do about tens of thousands of nameless and faceless children. In the modern world, it is an irrational moral calculation—rational economic man should care more for the many than the one. But an apparently irrational calculation becomes a rational moral choice in the ancient world of our evolved brains, where we care more for the one than the many, especially when the one is a proxy for those we evolved to care about—our immediate family, extended family, friends, community, and fellow in-group

members. The effect is deep and emotional, and it works. Indeed, after watching again that final scene from *About Schmidt*, I went online through World Vision's program to sponsor an eleven-year-old girl named Suada Isaku from Tirana, Albania, who lives in the rural farming district of Elbasan with her parents and sister struggling to survive on bread, vegetables, beans, and dairy products. My modest monthly donation, World Vision tells me, "will help provide Suada and her community with clean water and improved healthcare facilities. Your support will help create educational partnerships between parents and teachers to enhance students' education. Economic forums will help the community develop plans for growth."[55] An accompanying photograph with additional details about my sponsored child—she enjoys reading, helps at home with housework, likes to play ball games, and is in good health—reinforces my sense of attachment to her. A subsequent search on Google Earth promptly carried me through cyberspace to Suada's village, pulling on the heartstrings of my brain's dopamine reward networks, igniting my Middle Land propensity to connect to those near me, transforming a total stranger into honorary family through the power of markets, minds, and morals.

8

WHY MONEY CAN'T
BUY YOU HAPPINESS

In the 1979 cycling-cum-coming-of-age film *Breaking Away*, the Bloomington, Indiana, small-town recent high school grad Dave Stoller (Dennis Christopher) shaves his legs, drafts his Colnago bike behind eighteen-wheel trucks, and dreams of someday racing against the Italian pros. He finally gets his wish when Team Cinzano comes to town, only to discover that the elite cheat to win when a member of that team puts a pump through Dave's front fork and sends him into the pavement. Working at his father's used-car lot and struggling to hold on to his youthful ideals that hard work and honesty still reign, he offers a refund to a disgruntled customer, an act of rectification that stresses his more Machiavellian father to the cardiac breaking point. Recovering and asked if he feels lucky to be alive, Mr. Stoller (Paul Dooley) sardonically responds: "*No*, I don't feel lucky to be alive! I feel lucky I'm *not dead*. There's a difference."

Rationally speaking, there is no difference between being alive and being not dead; emotionally speaking, there is. Science can now tell us what that difference is. It turns out that such subjective evaluations, particularly how choices are framed (lucky to be alive versus lucky not dead), affect how we view the world, especially our happiness.

Happiness is a subjective state of well-being that depends on rel-
ative frames of reference, grounded in an evolved psychology that
finds meaning in the simple social pleasures and purposes of life.

In this simple statement we glean insight into one of the great para-
doxes of our time: traditional economics, grounded as it has been in rational choice theory, tells us that since people are selfish utility
maximizers, as they get wealthier they should get happier. We now have
the data to test that hypothesis and the unequivocal results are in—the
hypothesis is false. People are no happier today than they were in 1950,
despite the fact that by almost any imaginable material measure, life is
better today than it was then.

In fact, one of the most startling discoveries in the history of economics is that there is a disconnect between prosperity and happiness. Ever
since the time of Jeremy Bentham and the rise of the philosophy of utilitarianism in the nineteenth century, there has been a deeply held belief
among economists, politicians, and policy makers that an increase in utility would bring about an aggregate increase in national happiness; that is,
"the greatest happiness for the greatest number." Bentham even devised a
"hedonic calculus" to measure happiness, denoting "seven circumstances" by which "the value of a pleasure or a pain is considered": *purity*
("the chance it has of not being followed by sensations of the opposite
kind"); *intensity* (the strength, force, or power of the pleasure); *propin-
quity* (the proximity in time or place of the pleasure); *certainty* (the sure-
ness of the pleasure); *fecundity* ("the chance it has of being followed by
sensations of the same kind"); *extent* ("the number of persons to whom it
extends"); and *duration* (the length of time the pleasure will last).

Applied to society as a whole, Bentham instructs us to "Take an account of the *number* of persons whose interests appear to be concerned;
and repeat the above process with respect to each. *Sum up* the numbers
expressive of the degrees of *good* tendency, which the act has, with respect to each individual, in regard to whom the tendency of it is *good*
upon the whole: do this again with respect to each individual, in regard
to whom the tendency of it is *good* upon the whole: do this again with respect to each individual, in regard to whom the tendency of it is *bad*
upon the whole. Take the *balance*; which, if on the side of *pleasure*, will

give the general *good tendency* of the act, with respect to the total number or community of individuals concerned; if on the side of pain, the general *evil tendency*, with respect to the same community."[1]

Yeah, right. Let's see. There are three hundred million Americans, and we're going to establish economic, political, and social policy by conducting a national survey and then employing the hedonic calculus? Besides the logistical impossibility of such a task, the research on happiness—like that in behavioral economics—reveals that our feelings about life, positive and negative, are very much influenced by subjective states of mind and relative comparisons to others, thereby obviating any such attempts at establishing objective norms of happiness based on some economic measure of utility.

A short explanation for this disconnect between economics and happiness may be gleaned from the title of this chapter, in which we will explore the deeper evolutionary reasons for why money can't buy you happiness but family, friends, and purpose can.

✦ ✦ ✦

Over the past fifty years, standards of living have risen dramatically. As computed in 1996 dollars, the 1950 per capita Real Gross Domestic Product was $11,087, compared to the 2000 figure of $34,365.[2] That is a remarkable 300 percent increase in comparable dollars in only half a century. Not only is the average person absolutely richer than before, more are moving up the economic hierarchy. Way up. In 2000, one in four Americans earned at least $75,000 a year, putting them in the upper middle class, compared to a century before, in 1890, when only 1 percent earned the equivalent of that figure. That is a twenty-five-fold expansion of the upper middle class, redrawing class boundaries and redefining what it means to be average. And rich. Since 1980, the percentage of people earning $100,000 or more per year, in today's dollars, has doubled. And what we can buy with that money has also risen significantly. A McDonald's cheeseburger cost thirty minutes of work in the 1950s, three minutes of work today—an order of magnitude difference. In 2002, Americans bought 50 percent more health care coverage per person than they did in 1982.

We also have more SKUs in the form of SUVs, DVDs, PCs, TVs, designer clothes, name-brand jewelry and watches, stereos and home appliances, and electronic gadgets of diverse kinds. The homes in which we keep all of our goodies have doubled in size in just the past half century, from about 1,100 square feet in the 1950s to over 2,200 square feet today; 95 percent of these homes have central heating, compared to 15 percent a century ago, and 78 percent have air-conditioning, compared to the percent in our grandparents' generation—zip. According to the U.S. Census Bureau, in 2000, the median value of homes was $119,600, which is 18 percent more than the median value in 1990, and more than double the median value of $44,600 (expressed in 2000 dollars) in 1950.

That's not all. Crime is down while leisure is up. Most crime rates, almost everywhere, tumbled during the boom times of the 1990s. Homicides, for example, plummeted between 50 and 75 percent in such major cities as New York, Los Angeles, Boston, Baltimore, and San Diego. Domestic violence against women dropped 21 percent, while teen criminal acts fell by over 66 percent.

We now also enjoy a shorter workweek than our parents and grandparents, with the total hours of life spent working steadily declining for the past fifteen decades. In the mid-nineteenth century, for example, the average person invested 50 percent of his waking hours in the year working, compared to a mere 20 percent today. Fewer working hours translates into more leisure time. In 1880, the average American enjoyed just eleven hours per week of leisure time, compared to the forty hours per week average today. And those working environments are cleaner, safer, and more pleasant.

The good news is seemingly endless. Pollution is down, way down in many cities, such as my own Los Angeles. When I took up bicycle racing in 1979, the air was so bad that summer training rides had to be completed well before noon in order to avoid the pain caused by the fine particulate matter—dirt, dust, pollens, molds, ashes, soot, aerosols, carbon dioxide, sulfur dioxide, nitrogen oxides—becoming deeply embedded in your lungs. Today, I can ride practically any time of the day, nearly any day of the year, and feel no ill effects. My personal observation is backed by the data: during the 1980s, L.A. averaged 150 "health advisory" days per year and 50 "stage one" ozone alerts (we had to watch the evening news to determine our training ride for the next day). Thanks to the Clean Air Act and improved engine and fuel technologies, in the year 2000, there were only twenty health advisory days, and zero stage-one ozone alerts. And this example has been repeated throughout the United States over the past quarter century. Even though the number of cars has nearly doubled and the vehicle miles driven increased by nearly 150 percent, smog has diminished by a third, acid rain by two-thirds, airborne lead by 97 percent, and CFCs by virtually 100 percent. All of these improvements, and more (most notably in public health policy), have led to a doubling of the average life expectancy, from 41 in 1900 to the high 70s and low 80s in the modern Western world.[3]

Given these facts, and many more quantitative measures, it would be perfectly sane to decline a trip in a time machine to any point in the past if you had to actually live out your life there, which by comparison would be—*pace* Thomas Hobbes—nasty, brutish, and short. But reverse the time machine and transport someone to our time from the Middle Ages, the early modern period, or even the start of the Industrial Revolution, and we glean a perspective no prosperity metric could on just how far we have come since the world was lit only by fire. Back then, only the tiniest fraction of the population lived in relative comfort, while the masses toiled in squalor, lived in poverty, and expected half their children would die before adulthood and that they themselves would likely exit this world by their early forties. Most people never traveled beyond their hometown, and if they did it was on foot or by horse—Napoleon's troops moved no faster than Caesar's or Alexander's. No one—not even the wealthiest and most privileged aristocrats, kings, and clergy—enjoyed even the most quotidian technologies and public health benefits that we take for granted today. What would the average Adelaide or Anselm *then* make of the average Ashleigh or Allen *today* hurtling through the atmosphere at 400 miles per hour in a metal tube, crossing an entire continent in a few hours while listening to a Mozart concerto or a Shakespeare play on a wafer-thin pocket-size box through concert-hall-quality speakers the size of an ear canal, while writing on a portable laptop computer powered by rechargeable batteries and talking to someone thousands of miles away on another wafer-thin pocket-size box from 35,000 feet above the ground? *Past tense—future shock.*

✦ ✦ ✦

Despite all this wealth and prosperity, by all measures of Subjective Well-Being (SWB), people are no happier today than they were half a century ago. This has been called the *American Paradox*, the *Progress Paradox*, and the *Paradox of Choice*.[4] I call it the *Happiness Disconnect*. By whatever name, the phenomenon is real and needs an explanation through a closer examination of happiness itself.

The most common technique for measuring happiness is to just ask people. For example, the *Satisfaction with Life Scale* developed by Ed Diener presents people with five statements to which they can assign a number.[5] Try it yourself now:

1 = strongly disagree
2 = disagree
3 = slightly disagree
4 = neither agree nor disagree
5 = slightly agree
6 = agree
7 = strongly agree

_____*In most ways, my life is close to ideal.*
_____*The conditions of my life are excellent.*
_____*I am satisfied with my life.*
_____*So far, I have gotten the important things I want in life.*
_____*If I could live my life over, I would change almost nothing.*

Here is the scale to score yourself: *35–31 = extremely satisfied; 26–30 = satisfied; 21–25 = slightly satisfied; 20 = neutral; 15–19 = slightly dissatisfied; 10–14 = dissatisfied; 5–9 = extremely dissatisfied.*

How did you do? If you don't like that scale, there are others. For example: "Taking all things together, would you say you are *very happy, quite happy,* or *not very happy*?" When this question was asked of Americans in 1999, 38 percent said they were *very happy,* 53 percent *quite happy,* and only 9 percent *not very happy.* The percentages were nearly identical in England, and the figures remain largely the same whether the question is asked in writing or in person, and correlate highly with the evaluations of these same people by their friends and colleagues.[6] And these scales correlate highly with other happiness metrics.[7]

Such findings are also fairly consistent in cross-country comparisons.

In a comprehensive study of people from twelve European nations and the United States, for example, conducted every year between 1975 and 1992, a quarter of a million people were asked: "Taking all things together, how would you say things are these days—would you say you're *very satisfied, fairly satisfied, not very satisfied,* or *not at all satisfied* these days?" Closely matching the above study, overall 27 percent were *very satisfied*, 54 percent were *fairly satisfied*, 14 percent were *not very satisfied*, and 5 percent were *not at all satisfied*. (Adding a fourth category dilutes the percentages in each category.) Predictably, there were no significant income differences in happiness in the middle two categories of *fairly satisfied* and *not very satisfied*, but at the upper and lower ends there were differences: 33 percent of people in the highest-income quarter reported being *very satisfied* (with only 3 percent *not at all satisfied*), compared to 23 percent of those in the lowest-income quarter reporting that they are *very satisfied* (with 8 percent *not at all satisfied*). As Woody Allen quipped: "Money is better than poverty, if only for financial reasons."

But to the point I am making here, the percentages did not change much over the years, despite Europe's startling rise in economic prosperity. This disconnect can also be seen in America. To cite just one among hundreds of studies, a 1994 survey conducted by the Princeton Research Associates found that less than half of Americans feel they have enough money to lead satisfactory lives, and when asked, the vast majority of people said that they would like more money.[8] Well, who wouldn't? But compared to half a century ago we lead extremely satisfactory lives and we have way more wealth in absolute terms.

Tellingly, on the other end of the scale, the European study also found that two of life's greatest losses—unemployment and divorce—triggered significant dips in happiness. Only 16 percent of unemployed people were *very satisfied* compared to 39 percent who were either *not very satisfied* or *not at all satisfied*. Similarly, compared to 29 percent of married people who reported being *very satisfied*, only 19 percent of divorced people said that they were, whereas on the other end of the scale 17 percent of married people said that they were either *not very satisfied* or *not at all satisfied*, compared to 29 percent of divorced people.[9]

Arguably the most penetrating and most international study of happiness ever conducted is that of the World Values Survey that includes 250 questions that yield four hundred to eight hundred measurable variables. Over the past thirty years the survey shows that while people in certain pockets of the globe experience temporary increases or decreases in happiness, the average level of happiness has remained essentially unchanged.[10] Once average annual income is above $20,000 per annum in today's dollars—enough to put a roof over your head and provide three square meals a day for you and your family—more money does not bring more happiness.[11]

There are two reasons for this disconnect between money and happiness: genetics and relative value. Our genes account for roughly half of our predisposition to be happy or unhappy. That is, our temperament is the product of an interaction of our genetic heritage and our environmental circumstances (including the environments of our upbringing and those we choose today). In one massive twin study, scientists administered a fifty-item version of the Wilson-Patterson conservatism scale that measures different temperament types to 3,810 pairs of Australian twins and a forty-item Public Opinion Inventory to 825 British twin pairs. In both samples, subjects included identical and fraternal twins, allowing estimates of heritability based on the difference in correlations for individual survey items. On most psychological characteristics measured, the researchers found that about 40 percent of the variance in responses to the two surveys was heritable. Another 30 percent of the variance appeared to be explained by the shared family environment, leaving the remaining 30 percent of the variance to be explained by a combination of the nonshared environment and measurement error.[12]

Similarly, Niels Waller, Thomas Bouchard, and their colleagues in the Minnesota twins project measured numerous variables long thought to be primarily under the control of the environment, including temperament, personality, political attitudes, and religiosity. Studying 53 pairs of identical twins reared apart and 31 pairs of fraternal twins reared apart, the researchers found that the correlations between identical twins on most characteristics were typically double those for fraternal twins, a finding suggesting that genetic factors account for approximately half of the observed variance in their measures.[13]

If most personality characteristics are so highly heritable, what about happiness? Giving identical and nonidentical twins the Tellegen Multi-

Dimensional Personality Questionnaire as a measure of their Subjective Well-Being, Waller and Bouchard found that the correlation on happiness is 0.44 for identical twins and 0.08 for nonidentical twins, and that there was no significant difference in happiness between identical twins raised together and those raised apart. This makes it clear that it is not similar environments that make people similarly happy or unhappy, it is similar genes.[14]

These findings match our intuitions: we all know people who seem incessantly insouciant and energetically enthusiastic for life despite having suffered slings and arrows that would fell those with lesser dispositions. And we all know people who appear inescapably irascible and ill-tempered despite enjoying all the trappings of the good life. We marvel at the fortitude of the former and question the character of the latter. Given what we know about the power of genes, however, we would be well advised to temper our tendency to blame the environment for everything and recognize that nature matters, even if it is not all determining.

But what makes us happy or unhappy also tends to be relative to what other people have, and is not based on some absolute measure. This makes sense from the perspective of evolutionary economics. In our Paleolithic past we evolved in tiny communities of economic simplicity and relative equality, where happiness could not be found through wealth accumulation. The reason is that there was so little wealth to accumulate, and there was social pressure to redistribute what little wealth could be accrued to one individual or family. In folk economic terms, in Middle Land our senses and perceptions are geared for short-term assessments, direct comparisons, and relative social rankings. Who cares how people lived in previous generations about which our Paleolithic ancestors would have known next to nothing? What counts instead is the here and now, and what other people in our social group have compared to us.

Leaving genes aside, let's explore further the theory of happiness relativity.

✦ ✦ ✦

Would you rather earn $50,000 a year while other people make $25,000, or would you rather earn $100,000 a year while other people get $250,000? Prices of goods and services are the same. In other words, all other things being equal, would you rather make twice as much as other people or twice as much as yourself but less than half of other people? Surprisingly—stunningly, in fact—research shows that the majority of people select the first option: they would rather make twice as much as others even if that meant earning half as much as they could have. What a completely illogical decision! But as H. L. Mencken quipped, "A wealthy man is one who earns $100 a year more than his wife's sister's husband."

This study was conducted by the economists Sara Solnick and David Hemenway, who surveyed 257 students, faculty, and staff members at the Harvard School of Public Health. They made additional findings that bear witness to the subjective and relative nature of happiness. For example, when survey takers were asked if they would prefer to be the best-looking person in a community in which no one was particularly attractive, or absolutely hot in a land of 10s, people preferred the higher relative ranking even if it meant being absolutely less attractive. Likewise, most people preferred relative values when they were asked whether they wanted more education on an absolute level or less education but more than others, and whether they would prefer that their child be absolutely smart among a sea of brilliant kids or just relatively smart among a crowd of dunces.[15] In a broader and more representative sample of seven hundred randomly selected people ranging in age from eighteen to sixty-six in Sweden, researchers from Göteborg University found that the perceived value of income and cars is highly positional and depends on what others have in the society, whereas leisure time and car safety were less positional and more absolute.[16]

Wealth is even positionally relative to yourself and what you are accustomed to having. In another study people were asked, "What after-tax income for your family would you consider to be: *very bad, bad, insufficient, sufficient, good, very good*?" There was a strong positive correlation between income and perceived income need—the rich felt that they needed more income than the poor, to the tune of 40 cents on the dollar. That is, for every dollar increase in actual income there was an increase of 40 cents in "required income." So even though it is true that getting a raise this year makes you temporarily happy, next year you will have a new standard of need that is 40 percent higher that must be reached in order for you to feel the same level of satisfaction with your income.[17] How frustrating!—which is why economists call this effect the *hedonic treadmill*, a never-ending chase for a never-to-be-reached goal.

What these studies and others show is that when it comes to money, neither utility nor logic prevails. Another way to express the finding that people are willing to earn absolutely less if they can make relatively more is that we are willing to pay a price for relative rank and status, which is traded in a different form of currency—social capital. The economist Richard Thaler discovered a related irrationality he calls *regret aversion*, and he tested it by presenting people with a choice of being either Mr. A or Mr. B in the following scenario:

> Mr. A is waiting in line at a movie theater. When he gets to the ticket window he is told that as the one-hundred-thousandth customer of the theater, he has just won $100.
> Mr. B is waiting in line at a different theater. The man in front of him wins $1,000 for being the one-millionth customer of the theater. Mr. B wins $150.

Amazingly, most people said that they would prefer to be Mr. A. They were willing to forgo $50 in order to alleviate the feeling of regret that comes with not winning the thousand bucks.[18] That is, they were willing to pay $50 for regret therapy.

One solution to the happiness disconnect is to quit pursuing money as a means of attaining happiness. Another solution is to reframe the happiness question altogether. The latter is what the Emory University psychiatrist Gregory Berns thinks we should do.

Happiness is too often equated with pleasure, and it is the pursuit of pleasure that lands us on the hedonic treadmill. Because our sense of happiness tends to be based on positional and relative rankings compared to what others have, the pursuit of some absolute value that we believe will finally bring us happiness once we have achieved it paradoxically leads to misery when the goalposts keep moving. To get off the hedonic treadmill, Berns thinks that we need to reframe the question to involve *satisfaction* instead of happiness. "Satisfaction is an emotion that captures the uniquely human need to impart meaning to one's activities," Berns says.

"While you might find pleasure by happenstance—winning the lottery, possessing the genes for a sunny temperament, or having the luck not to live in poverty—satisfaction can arise only by the conscious decision to do something. And this makes all the difference in the world, because it is only your own actions for which you may take responsibility and credit."[19]

Harvard psychologist Daniel Gilbert offers a different solution to finding happiness: rather than proactively seek happiness, stumble upon it. Gilbert says that we are the only animal who thinks about the long-term future, and thus much of our happiness depends on projecting what *will* make us happy (instead of what actually does). Unfortunately, here we find yet another form of irrationality. Through a series of clever experiments, Gilbert has discovered that we are not very good at predicting what will make us happy or unhappy. He calls this *affective forecasting*. For example, in a comprehensive study involving six different experiments, Gilbert and his colleagues asked subjects to imagine how they would feel in a number of different scenarios that one could reasonably expect would trigger negative emotions, including the breakup of a romantic relationship, the failure to earn tenure, a defeat in a political election, negative feedback on one's personality, the death of a child, and a job rejection by a prospective employer. "Imagine that one morning your telephone rings and you find yourself speaking with the King of Sweden, who informs you in surprisingly good English that you have been selected as this year's recipient of a Nobel Prize," Gilbert suggests. "How would you feel, and how long would you feel that way?" Are you kidding? I'd be elated, and for a very long time. But wait, says, Gilbert. "Now imagine that the telephone call is from your college president, who regrets to inform you (in surprisingly good English) that the Board of Regents has dissolved your department, revoked your appointment, and stored your books in little cardboard boxes in the hallway. How would you feel, and how long would you feel that way?" Yikes, I'd be miserable, and for a very long time.

In neither scenario, says Gilbert, would I be elated or miserable for very long. The fact that I *think* I would be is called the *durability bias*, and we all suffer from it—incorrectly, as it turns out. "Research suggests that regardless of which call they received, their general level of happiness would return to baseline in relatively short order," Gilbert explains. "Common events typically influence people's subjective well-being for little more than a few months, and even uncommon events—such as losing a child in a car accident, being diagnosed with cancer, becoming paralyzed, or being sent to a concentration camp—seem to have less impact on long-term happiness than one might naïvely expect." In such situations we seem to experience *immune neglect*, where we neglect to consider the strength of our psychological immune systems to protect us against the pain of insult, defeat, regret, and loss. In all six experiments, for example, Gilbert and his colleagues found that "students, professors, voters, newspaper readers, test takers, and job seekers overestimated the duration of their affective reactions to romantic disappointments, career difficulties, political defeats, distressing news, clinical devaluations, and personal rejections." Why? Because of the durability bias and the failure to recognize the strength of their own psychological immune systems.[20]

On the positive end of the happiness scale, by contrast, most of us imagine that variety is the spice of life. But in an experiment in which subjects anticipated that they would prefer an assortment of snacks, when it actually came to eating the snacks week after week, subjects in the no-variety group said that they were more satisfied than the subjects in the variety group. "Wonderful things are especially wonderful the first time they happen," Gilbert explains, "but their wonderfulness wanes with repetition." Connecting psychology to economics to life, Gilbert notes with wry humor: "Psychologists call this habituation, economists call it declining marginal utility, and the rest of us call it marriage."[21]

On this last front, if you think that an array of sexual partners adds to the spice of marriage, you are probably mistaken. According to an exhaustive study published in *The Social Organization of Sexuality*, married people have more sex than singles . . . and more orgasms: 40 percent of monogamous married couples have sex twice a week, compared with only 25 percent of singles; married couples are more likely to have orgasms when they do have sex; and in almost all cases, men with one partner have more sex than men with multiple partners.[22] The historian of science Jennifer Michael Hecht emphasized this point in *The Happiness Myth*. Her deep-time perspective demonstrates just how time- and culture-dependent is all this happiness research. As she writes, "The basic modern assumptions about how to be happy are nonsense." Take sex. "If you are one of the many people who at some point in life feel sexually abnormal, note that a century ago a heterosexual married couple with cosmopolitan, secular values, having good sex three times a week, might well have felt shame and anxiety over it." By contrast, "A century ago, an average man who had not had sex in three years might have felt proud of his health and forbearance, and a woman might have praised herself for the health and happiness benefits of ten years of abstinence."[23]

The standards set for us by modern society distort our perceptions about what will actually bring us satisfaction because we evolved in an environment radically different from the one in which we now live.[24] Take location. Everyone thinks that where you live plays a sizable role in how happy you are. Living in California, for example, would on average surely make people happier than living in, say, Ohio. In fact, a study found that both Ohioans and Californians believe that living in California makes people happier, but in reality, when you compare data on measures of Subjective Well-Being between people from both states, there is no difference. Californians and Ohioans are equally happy.[25] In a related study, incoming college students were asked how they felt about their soon-to-be-announced dorm room assignments. The researchers found that students vastly overestimated the impact that the housing assignment would have on their happiness.[26] That is, "location, location, location" is a myth when it comes to the actual effect on people's Subjective Well-Being.

The proximate reason for these effects is that the relative standards to which we are so hypersensitive derail us from focusing on those things that will, in fact, bring us satisfaction and happiness. The ultimate reason for these effects is that in our evolutionary environment, friendships, love, children, hard work on behalf of one's family and community, and socializing with fellow group members were the primary road to psychological satisfaction. We shall conclude this chapter by looking at the proximate reasons (neuroscience) and ultimate reasons (evolution) for happiness.

◆ ◆ ◆

How did an emotional system like happiness and unhappiness evolve, and from where in the brain do such emotions arise? EEG measures of the brains of subjects who report positive emotions or view a funny video clip show increased activity in the left front cortex, whereas negative emotions and unpleasant video clips show increased activity in the right front cortex.[27] PET scans and fMRI scans of subjects who are exposed to a photograph of a cute little baby show increased activity in the same left front cortex area, whereas subjects who are exposed to a photograph of a grotesquely deformed little baby show increased activity in the same right front cortex area.[28] You can even reverse the process, bombarding the left front cortex of the brain with a strong magnetic field to elicit a positive mood, and the right front cortex to elicit a negative mood.[29]

An evolutionary connection can be seen in a 2007 study on tail wagging in dogs conducted by the Italian neuroscientist Giorgio Vallortigara and his veterinarian colleagues at the University of Bari, who found that dogs wag their tails more to the right when they are feeling positive and more to the left when they are feeling negative. In an experiment involving thirty mixed-breed dogs, each was placed in a cage equipped with cameras that measured the asymmetrical bias (left or right) of tail wagging while they were exposed to four stimuli: their owner, an unfamiliar human, a cat, and an unfamiliar dominant dog. Dogs' owners elicited a strong right bias in tail wagging, unfamiliar humans and a cat triggered a slight right bias, but the unfamiliar dominant dog (a large Belgian shepherd Malinois) elicited a strong left bias in tail wagging. So the muscles in the right side of the tail were responding to positive emotions in the brain, while the muscles in the left side of the tail were responding to negative emotions in the brain. Why? Because, as we saw in people, the left brain is associated with positive emotions such as love, attachment, bonding, and safety, and similar research on other animals shows that birds, fish, and even frogs show left-brain/right-brain differences in approach-avoidance behavior, with the left brain associated with positive approach feelings and the right brain associated with negative avoidance feelings. Since the left brain controls the right side of the body and the right brain controls the left side of the body, the nerve signals are crossing the midline of the body and causing, in this example, the dog's tail to wag more to the right when its left brain is experiencing a positive emotion. One final evolutionary connection can be found in chimpanzee brains, which are equally asymmetrical as human brains; when chimps are experiencing negative emotions, they tend to scratch themselves on the left side of their bodies, and left-handed chimps, whose right brain is dominant, tend to be more fearful of novel stimuli than right-handed chimps.[30]

In keeping with the behavior genetics findings that as much as half of the variation in temperament between people can be accounted for by our genes, when people who are naturally happy are exposed to pleasant video clips they experience a greater gain in happiness than those people who are less naturally happy, and those whose left frontal cortex is naturally more active than their right respond more favorably to positive scenes.[31] Psychologist and twins researcher David Lykken even suggests that we have a "set-point for happiness" that is initially set by our genes and then tweaked and modified by our environment.[32] This would explain why lottery winners and recently incarcerated prisoners both return to their original set-points of happiness within a year of their life-changing event.

If there are neural networks and brain modules specifically involved in emotions, they must be there for a reason. That is, they must have some adaptive function. What is the purpose of emotions? Emotions interact with our cognitive thought processes to guide our behaviors toward the goal of survival and reproduction. At low levels of stimulation, emotions appear to play an advisory role, carrying additional information to the decision-making process along with inputs from higher order cortical regions of the brain. At medium levels of stimulation, conflicts can arise between high-road reason centers and low-road emotion centers. At high levels of stimulation, low-road emotions can so overrun high-road cognitive processes that people can no longer reason their way to a decision and report feeling "out of control" or "acting against their own self-interest."[33]

Fear is another emotion with an obvious adaptive purpose. The Norwegian anthropologist Björn Grinde notes how quickly a dangerous situation can switch from being a positive emotion of the thrill of taking a risk to a negative emotion of fear of death. A mountain climber himself, Grinde notes what happens when a rock climber loses his grip. "A scare is typically perceived as pleasant if the individual retains control of the situation, while unpleasant if the situation gets out of control, because these two modes of experiencing fright serve different biological purposes: The brain is designed to induce us to take some chances, otherwise we would never have laid down a large prey or ventured into uncharted land; but it is also designed to stop us from causing harm to ourselves, that is, to avoid hazards. The 'adrenaline kick' associated with climbing a mountain or riding a roller coaster may feel good, presumably because it improves the chance of survival if voluntarily encountered dangerous situations induce a positive mood and a high self-esteem. At the moment one loses the grip on the mountain, the unpleasant sensations devoted to harm avoidance kicks in."[34]

In like manner, a little bit of hunger may be perceived as pleasant, as it motivates you to seek and find food, thereby shifting needs into wants. Of course, too much hunger becomes an unpleasant emotion when it goes unmet. Here we see that emotions act as a feedback mechanism to alert the brain when the body is out of balance, or deviates from *homeostasis*. Think of the process as an emotional thermostat. When our bodies are low on energy we feel hungry, and that emotion is triggered by a number of internal and external feedback cues, such as shrinking or distension of the stomach, elevated or reduced blood glucose levels, and the sight or smell of food. Similarly, when our core body temperature deviates above or below the 98.6 degree Fahrenheit set-point, systems kick in to correct the imbalance—sweating to cool the body, shivering to warm it up. Departing from the set-point of a homeostatic system feels bad, and this negative emotion motivates the animal to take action to correct the imbalance. Moving the out-of-balance system back toward homeostasis feels good, and behaviors that feel good tend to be repeated. Thus it is that in our need to maintain homeostasis, our emotions direct us to avoid pain and pursue pleasure, or to avert unhappiness and seek happiness.

Even deeper emotions such as depression take on new meaning in an evolutionary context. The symptoms of depression—restlessness, agitation, loss of appetite, disturbed sleep, impaired concentration, and loss of motivation—may not be signs of an illness; rather, they may represent an adaptive response to prod you into doing something different in your life. The evolutionary psychologists Peggy La Cerra and Roger Bingham make this point in their book *The Origin of Minds*. "Because behavior is so enormously expensive energetically, the best thing a person in this situation can do is to stop what he has been doing, reconfigure his life, and try to formulate a more viable trajectory into the future." What would have been the evolutionary pressures to lead to such a behavioral response? "If you were an ancestral human who was being exploited by another individual or group of individuals, a complete behavior shutdown could abruptly force a renegotiation of the inequitable social relationship." Even in the modern world, depression "serves as a wake-up call, prodding people to abandon dead-end jobs and relationships."[35] Sure enough, the same research that illuminates what triggers an increase in happiness also reveals what brings unhappiness: *divorce, unemployment, poverty, loss of status,* and *depression*. Happiness and unhappiness are emotions, and we know that emotions have powerful effects on people's perceptions and behavior. Here again, balance is the key—too much unhappiness can undo what a little can help. Studies show, for example, that excessively sad people tend to recall sad memories, which in turn increases their sadness, leading to even more sad memory recall, and so on in a negative spiral. Sad people also tend to feel more threatened by others and situations, whereas the emotion of anger makes people more likely to take risks.[36]

In the mind of the market, a homeostatic model of emotions goes a long way toward explaining why so many economic values are relative, subjective, and dependent on comparisons to other values. Homeostatic systems are finely tuned to relative changes in stimuli rather than their absolute levels. The behavioral economist Colin Camerer makes this point when he notes that "rather than viewing pleasure as the goal of human behavior, a more realistic account would view pleasure as a homeostatic cue—an informational signal." For example, he says, "Neural sensitivity to change is probably important in explaining why the evaluation of risky gambles depends on a reference point which encodes whether an outcome is a gain or a loss, why self-reported happiness (and behavioral indicators like suicide) depend on changes in income and wealth, rather than levels, and why violations of expectations trigger powerful emotional responses."[37] This also helps explain why sometimes too much thinking and not enough feeling can lead to bad choices. Experiments show, for example, that when subjects are given a choice between desired objects, those who were asked to give their reasons for why they made the choice that they did ended up less happy with their choices than subjects who were just asked to choose based on what felt right to them.[38]

Yet there is more to the story than simply doing what "feels right." As Bentham showed, the utilitarian philosophy of seeking the greatest happiness for the greatest number does not simply involve the pursuit of pure hedonic pleasure, as in "If it feels good, do it." Factoring in his seven variables, as often as not we should forgo immediate pleasures of the flesh in favor of more lasting and long-term pleasures of the mind and morals. What are the neural correlates of Bentham's system of balanced pleasures? The neuroscientist Ken Berridge makes a useful distinction between brain systems responsible for pleasure and pain (the *liking system*) and brain systems responsible for motivation (the *wanting system*).[39] We may *want* something that offers only a modicum of pure pleasure more than we *like* something that promises a high degree of untainted delight. Berridge has shown, for example, that you can alter a rat's willingness to work for food simply by changing the conditions in the environment but without changing how much pleasure the food brings the rat.

Berridge even speculates that drug addicts may be motivated by *wanting* the drug even if they do not actually *like* taking it. Or it may be that addictive drugs do not feel good so much as they decrease the anxiety of wanting the high; that is, they are negatively reinforcing, or they remove an aversive stimulation. In this model, the removal of the want is what is motivating. The same may be true for Obsessive-Compulsive Disorder—the compulsive act lowers the anxiety that builds up from the obsessive thoughts. OCD may be part of a wanting system rather than a liking system. Russell Poldrack made a similar point about the experiments I discussed in the previous chapter in which rats pressed a bar that sent an electrical signal to their nucleus accumbens in the ventral striatum, which scientists have long assumed must "feel good" because they do it until they drop. Since stimulating the nucleus accumbens leads to an increase in dopamine, does that make dopamine the "feel good" brain drug, I wondered? Poldrack responded, "There is a distinction between 'liking' and 'wanting,' and ventral striatum self-stimulation may be more about the latter. Certainly dopamine is more about wanting than liking."[40]

In this sense, emotions evolved a dual system of liking and wanting—of reward seeking and punishment avoidance—to help us make choices with a proximate goal of maintaining balance and homeostasis and an ultimate goal of survival and reproduction.

✦ ✦ ✦

Happiness is a subjective state of well-being that depends on relative frames of reference, grounded in an evolved psychology that finds meaning in the simple social pleasures and purposes of life. To cut to the chase, two decades of extensive research on Subjective Well-Being from around the world has revealed what brings most people the most happiness: *social bonds* (marriage, friendships, social circle), *trust in people* (friends, family, and strangers), *trust in society* (the economy, justice, government), *religion and spirituality* (prayer, meditation, positive psychology), and *prosocial behavior* (helping others, aiding the needy, volunteering). The reason why money can't buy you happiness, but family, friends, and purpose can, is that these things are tapping into the environment of our evolutionary adaptation—the Middle Land of our Paleolithic past—where social bonds, trust, and helping others in the group were critically important to the survival of a diminutive social primate species struggling to endure and compete against other social primate groups, against predators, and against an unforgiving environment.

In the previous chapter, I argued that evolution designed us to value kin over nonkin, friends over strangers, and in-group members over out-group members. The emotional differences we feel about these various categories of people reflect a rational calculation conducted over the evolutionary eons. Emotions are proxies for getting us to act in ways that lead to an increase in reproductive success. Just as we do not need to compute the caloric value of food, the genetic value of mates, or the moral value of others, positive emotions like attachment and love guide us toward committing to a monogamous relationship in order to better ensure the survival of our genes into future generations, and negative emotions such as jealousy clue us in to the dangers of investing in someone else's genes. We don't need to do these calculations because evolution has done them for us and we have only to let our emotions guide us.

Negative emotions, by contrast, evolved as adaptations to threats faced by our Paleolithic ancestors. *Anger* leads us to strike out, fight back, and defend ourselves against danger. *Fear* causes us to pull back, retreat, and escape from risks. *Disgust* directs us to expel that which is bad for us. Computing the odds of danger in any given situation takes too long. We need to react instantly. Positive emotions, by contrast, have more long-term evolutionary benefits. Barbara Fredrickson, a psychologist who directs the Positive Emotions and Psychophysiology Laboratory at the University of Michigan, suggests that positive emotions help us build enduring personal resources, including *intellectual resources* (problem-solving skills, acquiring new information), *physical resources* (coordination, strength, health), *psychological resources* (resilience, optimism, sense of identity), and especially *social resources* (reinforce old bonds, create new bonds). Positive emotions, says Fredrickson, "can transform people for the better, making them more optimistic, resilient, and socially connected. Indeed, this insight might solve the evolutionary mystery of positive emotions: simply by experiencing positive emotions, our ancestors would have naturally accrued more personal resources. And when later faced with threats to life or limb, these greater resources translated into greater odds of survival and greater odds of living long enough to reproduce."[41]

Play is a wonderful example of a positive emotion with an adaptive purpose. The psychologist and evolutionary biologist Gordon Burghardt, in his engaging book *The Genesis of Animal Play*, explains how play elicits a sense of joy because it leads to a number of adaptive characteristics essential to survival, such as improved coordination, enhanced cardio-vascular strength and endurance, integration of the senses and nervous system, honed ability to react appropriately to others, and determination of social status and gender roles. Play also involves practicing prey capture, practicing parenting skills, learning what objects do and how other animals behave, gaining a source of novel behavioral responses to the environment, honing problem-solving abilities, strengthening psychological resiliency, and most notably, developing social bonds with playmates. Play, says Burghardt, has "an important role in the behavioral, social, emotional, cognitive, physiological, and developmental realms in the lives of many animals, including people."[42]

What I want to argue here is that happiness is an emotion no different in principle from the other emotions, which evolved as proxies to guide how we behave toward others. Evolution designed us to care the most about, and thus feel happiest toward, meeting our basic needs: procreation and sex, bonding and attachment to spouse and children, affiliation and affection for extended family members, and empathetic concern for our fellow group members and those in our immediate community. To expand on my earlier definition, *happiness is an evolved emotion that guides us to find meaning in the simple social pleasures of interacting with our immediate family and extended family, friends, and social circle, and to direct us to find joy in the meaningful purposes of life that most directly involve helping ourselves, our family, our friends, and our community.*

Countless studies on what makes people happy support this evolutionary explanation. Marriage, for example, is one of the strongest predictors of happiness. As seen in data collected by the National Opinion Research Center at the University of Chicago, married people are happier than singles and divorced people. Since marriage is one of the deepest social bonds any two people can form, this makes sense in a context of emotional evolution where social participation and social support are so vital to human survival and well-being. This also explains why divorce and the death of a spouse or close friend can be so devastating.

Broadening the social circle, consider the results of one extensive survey conducted on a thousand workers in Texas, who were asked to divide the previous day into fifteen episodes and tell what they were doing in each one, whom they were doing it with, and how they felt about it as measured by twelve different dimensions that were then combined into a single figure for satisfaction.[43] People most like to have sex, socialize, relax, eat meals, and then exercise, and most like to participate in these activities with friends, relatives, spouse, children, and then coworkers. It is clear that people most like doing things with other people, especially if they are people to whom we are closely attached (friends and spouse) or to whom we are closely related (family).

Consider as well the research by psychologists David Myers and Ed Diener on four personality traits that correlate highly with happiness:

1. *High self-esteem*. Happy people like themselves, and believe themselves to be healthier, more intelligent, more ethical, less prejudiced, and better able to get along with other people.
2. *Personal control*. Happy people have the freedom to control their own lives, their choices and decisions, and therefore life outcomes. People lacking such freedom typically suffer lower levels of happiness, morale, and even health, such as prisoners, long-term-care patients, citizens of impoverished countries, and citizens of totalitarian regimes.
3. *Optimism*. Happy people view the world in a more positive manner, tending to see the good in others and in events.
4. *Extroversion*. Happy people are personally outgoing and socially gregarious. They like being around other people, and that in turn brings them more social contacts and opportunities for warm and caring relationships.[44]

Of course, it is not clear which direction the causal arrow points. Does being happy make you more extroverted, optimistic, confident, and in control, or are people with these characteristics happier? It appears to be some combination of both personality traits and emotional states, interacting in a feedback loop. Extroverted people tend to be more sociable and meet more people, which increases their social sense of well-being, and that elevated positive emotion increases their self-esteem and optimism and makes them feel more in control, and these states in turn increase their happiness and make them more sociable, and round and round it goes in yet another example of a feedback loop.

Regardless of how, precisely, such traits and states interact, my point here is that being social is integral to all aspects of our lives, including our Subjective Well-Being. Even the findings that religiously active people report higher levels of happiness are probably related to our social nature. One Gallup survey, for example, found that people who rated themselves as highly religious were also twice as likely to call themselves very happy than those who rated themselves as low in religious commitment. A sixteen-nation international study of 166,000 people found a strong correlation between reported happiness and life satisfaction and the strength of their religious affiliations and church attendance. Myers and Diener suggest "that religious affiliation entails greater social support and hopefulness."[45] Why should this be? What is it about religion that makes people happy?

It is because of our evolved social and moral natures, and there are few social institutions that are more tightly knit than religion, which is why the theory of religion that I present in *How We Believe* includes a strong social component. Religion is a social institution that evolved as an integral mechanism of human culture to encourage altruism, reciprocal altruism, and indirect altruism, and to reveal the level of commitment to cooperate and reciprocate among members of a community. That is, religion evolved as a social structure that enforced the rules of social interactions before there were such institutions as the state or such concepts as laws and rights. We would do well to remember that the history of the modern nation-state with constitutional rights and protection of basic human freedoms can be measured in mere centuries, whereas humans evolved as social primates over the course of millions of years, and human culture itself dates back at least a hundred thousand years. The principal social structure available to facilitate cooperation and goodwill was some form of religion. Religion is an organized institution with rules and morals, with the hierarchical structure so necessary for social primates, and with belief in a higher power to enforce the rules and punish transgressors.[46] This would explain its ubiquity. According to the Oxford *World Christian Encyclopedia*, of the earth's 6.1 billion humans in 2001, 5.1 billion of them—84 percent—belonged to some form of organized religion. There are more than ten thousand distinct religions, each one of which can be further subdivided and classified. In Christianity alone, for example, there are no fewer than 33,820 different denominations.[47] If religion were not so important to the lives of people, we would not see such numbers. People do not attend religious services just to hear a sermon on the creation of the universe or the meaning of a religious doctrine. For whatever other reasons people participate in such religious endeavors, the social aspects involved are not trivial. And it is that sociality that makes people happy.

◆ ◆ ◆

One final component to happiness is finding meaning and purpose in life. The Hungarian psychologist Mihaly Csikszentmihalyi (pronounced *chick-sent-me-high*, but "Mike" to his friends), for example, discovered the concept of *flow*, or the experience of being totally absorbed in meaningful tasks that fully engage the person. As he explained it to me when I interviewed him, "Flow is a state of total involvement and intention, where you have clear goals with immediate feedback, where you lose a sense of time, you lose a sense of your social self such that you are not self-conscious and you are not afraid of being embarrassed; it is finding the right balance between being challenged enough to grow but not being too challenged that failure is inevitable, and it is a sense that the activity is intrinsically rewarding such that external rewards are not necessary."[48]

I call this the *lottery test*. How would your life change if you won the lottery? Most people say that they would quit their jobs and change their lives dramatically. This means that they are not presently living the life they want; they are not doing what fulfills them; they are not in flow. How many people are in or out of flow? According to Mike, flow fits the normal distribution, where "10–12 percent never experience flow, 10–12 percent experience flow every day, and the rest are in between." A husky medium-sized man with the rounded facial features that reveal his eastern European heritage, Mike's thick accent and bushy eyebrows bring to mind that other genius from his country, Edward Teller. Mike's childhood growing up in Nazi-occupied Hungary during the Second World War and Communist-run Hungary after the war help explain why understanding concepts like happiness and flow became so important to him. "I was dismayed to find out that grownups had no idea what was going on, and were helpless to extricate themselves from the mess they had created," he explained. "I resolved to figure out how one could live a better life. I discovered psychology through the writings of Carl G. Jung, and thought that perhaps this was the best way to understand behavior and history." Six decades on, he is still searching. We are all still searching. My answer that follows is necessarily personal and subjective (even while being grounded in science), but then so too is happiness.

How and why would evolution have created in us a sense of purpose to the point where we would find happiness in pursuing it? In keeping with my thesis that moral sentiments and happiness are evolved emotions with an adaptive function, we can consider purpose to be an evolved emotion—a psychological desire to accomplish a goal—that developed out of behaviors selected for because they were good for the individual or for the group. Although cultures may differ on what behaviors are defined as purposeful, the desire to behave in purposeful ways is an evolved trait. Purpose is in our nature. Evolution gave us a *Purpose-Driven Life*. How we define our purpose-driven lives may be personal, but the deep sense of wanting a purpose is an evolved characteristic. How?

Life began with the most basic purpose of all: survival and reproduction. For 3.5 billion years, organisms have survived and reproduced in a lineal descent from the Precambrian to us, an unbroken continuity from bacteria to big brains that has endured countless terrestrial and extraterrestrial assaults and at least six mass extinctions—at any point along the journey, life on earth could have been easily snuffed out, never to arise again. So the ultimate purpose of our life is to survive and reproduce in order to keep the chain unbroken. As Charles Darwin wrote in the penultimate paragraph of his 1859 masterpiece, *On the Origin of Species*: "When I view all beings not as special creations, but as the lineal descendants of some few beings which lived long before the first bed of the Silurian system was deposited, they seem to me to become ennobled."

Feeling ennobled is a pleasurable emotion that arises out of this deepest sense of purpose. Although there are countless activities people engage in to satisfy this deep-seated need, the research shows that there are at least four means by which we can bootstrap ourselves toward happiness through purposeful action. These include:

1. *Deep love and family commitment*: Bonding and attachment to others increases one's circle of sentiments and corresponding sense of purpose to care about others as much as, if not more than, oneself.
2. *Meaningful work and career*: The sense of purpose derived from discovering one's passion for work drives people to achieve goals so far beyond their own needs that they lift all of us to a higher plane, either directly through the derivatives of the work itself, or indirectly through inspiration and role modeling.

3. *Social and political involvement*: As a social species we have an obligation to our community and our society to participate in the process of determining how best to live together.
4. *Transcendency and spirituality*: This capacity, unique to our species, includes aesthetic appreciation, spiritual reflection, and transcendent contemplation through a variety of expressions such as art, music, dance, exercise, meditation, prayer, quiet contemplation, and religious reverie, connecting us on the deepest level with that which is outside of ourselves.

To that end, I suggest three principles that tie together the findings on happiness, liberty, and purpose:

1. *The Happiness Principle*: It is a higher moral principle to always seek happiness with someone else's happiness in mind, and never seek happiness when it leads to someone else's unhappiness.
2. *The Liberty Principle*: It is a higher moral principle to always seek liberty with someone else's liberty in mind, and never seek liberty when it leads to someone else's loss of liberty.
3. *The Purpose Principle*: It is a higher moral principle to pursue purposeful goals with someone else's purposeful goals in mind, and never pursue a purpose when it leads to someone else's loss of purpose.

Evolution created in us a basic drive of purpose, but higher moral purposes are learned. To reach the highest levels of moral purpose that concern society, the species, and the biosphere—especially with people who are not related to us, are not in our social group, or belong to other groups on other continents whom we shall never meet—requires volitional action and a social conscience. This chapter—indeed, this entire book—is not just descriptive of the way the world works, but it is also prescriptive of the way the world *should* work. That is, this is an exercise in raising our consciousness to higher economic, political, and moral levels. As one of the great consciousness-raisers of the twentieth century, Helen Keller, wrote in a 1933 *Home Magazine* article entitled "The Simplest Way to Be Happy":

I know no study that will take you nearer the way to happiness than the study of nature—and I include in the study of nature not only things and their forces, but also mankind and their ways, and the moulding of the affections and the will into an earnest desire not only to be happy, but to create happiness. It all comes to this: the simplest way to be happy is to do good.[49]

9

TRUST WITH CREDIT VERIFICATION

In John Ford's classic 1962 film, *The Man Who Shot Liberty Valance*, a clash of moralities unfolds in the Wild West frontier town of Shinbone, Arizona. There in the dusty streets and ramshackle buildings two self-contained and internally consistent moral codes come into conflict. One is the *Cowboy Ethic*, in which trust is established through courage, loyalty, and personal allegiance to friends and family, and disputes are settled and justice is served between individuals who have taken the law into their own hands. The other is the *Law Ethic*, in which trust is established through the transparent and mutually agreed upon rule of law, and disputes are settled and justice is served between all members of the society who, by virtue of living there, have tacitly agreed to obey the rules. Only one of these codes can prevail.

In the film we find a fictional analogue to the problems we will consider in this and the next chapter: first, how trust is established between strangers on a psychological level through mutually desired encounters and reciprocally beneficial exchanges that turn total strangers into honorary friends; and second, how trust is established on a societal level through institutions that encourage cooperative and prosocial behavior and enforce the rules that ensure the fairness and justice of such encounters and exchanges. In order to live in a large modern society we need both psychological trust and social technologies that reinforce trust, and this is understood through the study of how the mind works and how social institutions operate. During those long-gone millennia when hunter-gatherers morphed into consumer-traders, there had to be a conflict between these systems, as the Cowboy Ethic inexorably gave way to the Law Ethic.

In *The Man Who Shot Liberty Valance*, the Cowboy Ethic is represented by two people, one good and the other evil. John Wayne's character, Tom Doniphon, is a fiercely loyal and deeply honest gunslinger duty-bound to enforce justice on his own terms through the power of his presence backed by the gun on his hip. Lee Marvin's title character, Liberty Valance, is a coarse and unkempt highwayman whose unruly behavior provokes fights with the locals, most of whom fear and loathe him. The Law Ethic is represented by Jimmy Stewart's improbably named character, Ransom Stoddard, an attorney hell-bent on seeing his beloved Shinbone make the transition from cowboy justice to the rule of law. John Ford opens his film at the end of the story: the funeral of Tom Doniphon, attended by an elderly Stoddard, who is swamped by reporters inquiring why the now-distinguished U.S. senator would bother returning to his native town for the memorial service of a down-and-out gunfighter.

When they were coming of age in this territory just slightly out of reach of the law, Stoddard and Doniphon were of radically different minds about how justice should be served, each believing that the other's strategy is either outdated (Doniphon's gun) or naïve (Stoddard's law). Despite this difference, or perhaps because of it, they become faithful friends, both believing that in the end justice must prevail. When Valance arrives it is clear that he respects only Doniphon, because they share the Cowboy Ethic that men settle their disputes between themselves. As Doniphon boasts, "Liberty Valance is the toughest man south of the Picketwire—next to me." But Valance's disdain for the milksop Stoddard and his naïve notions about the effectiveness of the law knows no bounds. Entering a restaurant where Stoddard is dining, for example, Valance berates him, taunts him, and finally trips the waiter, sending Stoddard's dinner to the floor. As Stoddard meekly tries to avoid a confrontation, Doniphon enters and stares down Valance, who snaps back, "You lookin' for trouble, Doniphon?" In his inimitable John Wayne drawl, Doniphon responds, "You aimin' to help me find some?" Valance caves to Doniphon's challenge and scurries out of the restaurant. "Well, now; what do you supposed caused him to leave?" Doniphon wonders rhetorically. The sarcastic response from a patron in reference to the impotency of Stoddard's philosophy reveals which ethic is still dominant: "Why, it was the specter of law and order rising from the gravy and the mashed potatoes."

Despite Valance's relentless taunting, Stoddard holds to his belief that until Valance is caught doing something illegal there can be no justice. When Doniphon tells Stoddard, "You better start packin' a handgun," Stoddard, rejoins, "I don't want to kill him. I just want to put him in jail." At long last, however, Stoddard can take the derision no more, so he decides to take Doniphon's advice that "out here a man settles his own problems" and turns to him for gunfighting lessons. When Valance challenges Stoddard to a duel, the overconfident naïf accepts and a late-night showdown ensues. On a darkened street, the two men square off. Stoddard is trembling in fear while Valance mocks and scorns him, shooting first too high and then too low. When Valance takes aim to kill, Stoddard shakily draws his weapon and discharges it, and Valance collapses in a heap. Having felled one of the toughest guns in the West, Stoddard becomes a local hero, building that image into political capital and working his way up from town politics to a distinguished career in the U.S. Senate.

So it would appear that the Law Ethic prevailed, literally and figuratively through the flash of a gun. Moral conflict resolved.

But as the film continues we learn that sometime after the gunfight, Stoddard discovered that he was not the fatal shooter. The man who shot Liberty Valance was Tom Doniphon. Knowing that Stoddard was no match for Valance, Doniphon lurked in the shadows, fingering a rifle, which he engaged to kill Valance at the crucially timed moment when the two men drew their weapons. Holding to the Cowboy Ethic of loyalty and friendship, Doniphon takes the secret to his grave. When Stoddard finally reveals to a newspaper reporter the truth about who really shot Liberty Valance, the paper decides not to print the story because, in what has become one of the most memorable lines in filmic history, "When the legend becomes fact, print the legend."

The philosopher Patrick Grim, who called my attention to the film as a tale of moral conflict, notes that both Stoddard and Doniphon violated their principles, but they do so because this was the only means by which one moral code could displace the other.[1] By agreeing to a duel with Valance, Stoddard adopts a form of conflict resolution that he previously deemed illegal and immoral, and after discovering the truth about who really shot Valance, he chooses to live a lie of omission and then capitalizes on his unearned reputation for heroism. For his part, Doniphon violates his moral code by ambushing Valance instead of facing him man to man, and then hides the truth about what really happened, tacitly endorsing Stoddard's faux use of the Cowboy Ethic in order to help bring about the Law Ethic to town. In fact, both men violated their own codes of morality, and with ample irony, the only person who held true was the scurrilous Liberty Valance. But in the end, as Shinbone grew in size, the transition from one moral code to the other had to happen, and in this moral homily it was friendship and loyalty that facilitated the change. It was the psychology of trust between individuals that enabled a society of trust among the collective to come to fruition.

The fictional Shinbone embodies any small community in transition from an informal to a formal moral code and system of justice. As long as population numbers are low and everyone in a community is either related to one another or knows one another through regular interactions, the code of the cowboy can work relatively well to keep the peace and ensure trust and social stability. But when communities expand and population numbers increase, the opportunities for unchecked violations of such informal codes expand exponentially, requiring the creation of such social technologies as laws codes, courts, and constitutions. The transition from the informal rule of frontier justice found in premodern societies to the formal rule of law pervasive throughout modern democratic nations is a result of the creation of a myriad of social technologies and economic institutions.

The development of trust between individuals and among societies is a product of our evolved moral emotions and our personal encounters with others, including and especially various forms of reciprocity, exchange, and trade. Three lines of evidence link evolution to economics through trust and trade: primate politics, human cooperation, and cross-cultural exchange.

✦ ✦ ✦

Frans de Waal is no stranger to primate politics, both of the human and nonhuman kind. A professor of psychology at Emory University and director of the Living Links Center for the Advanced Study of Ape and Human Evolution, De Waal doesn't just teach about the psychology of primate politics, he monitors it through a wall-size window from his office high above a large chimpanzee enclosure; he also observes it among his primate colleagues in science. When I visited his lab, De Waal was investigating how primates resolve conflicts, punish transgressors, and learn to cooperate with one another.

De Waal studies one of the thorniest debates in all of science. To what extent are our behaviors determined by our genes versus our environment; to what degree do genes play a role in shaping our behavior, and to what extent were they selected for because of their adaptive function versus being later co-opted for a new purpose (an exaptation instead of adaptation)? De Waal says he follows the sage advice of the renowned evolutionary biologist George C. Williams that "adaptation is a special and onerous concept that should be used only when it is really necessary." De Waal has managed to strike a healthy balance between nature and nurture in his work, so as we settle into his glass-enclosed observation room I ask him why his view of evolution is so different from that of other evolutionary biologists.[2] "Sometimes I attribute my interest in aggression in primates to the fact that I am the fourth of six brothers. In most of the literature, aggression is presented as a problem that needs to be solved, whereas my experience is that a little bit of aggression is a normal part of human interactions." Since De Waal is Dutch, and Holland is a fairly crowded country by American standards, I inquire about the famous "behavioral sink" experiments connecting overcrowding and aggression that were conducted in the late 1950s by the ethologist John B. Calhoun. "I never believed the Calhoun experiments when I first read about them because Holland is a very crowded country, and in fact you don't see the behavior that Calhoun reported in his research. And when we actually reviewed the literature for a comprehensive study on population densities and murder rates of numerous countries, we found that there was no correlation whatsoever. Then we looked at different subcategories of countries and we found one correlation—for the Eastern bloc countries—where the murder rate was highest among the least populous! Russia is a good example."

Yet the overcrowding-aggression connection seems to make sense intuitively, if not scientifically. De Waal has a rejoinder. "I would argue that the intuitive correlation would be the opposite," De Waal says. "If you live closely together for many generations you are going to develop rules of conduct that prevent excessive aggression and violence. In Japan, Holland, Bangladesh, and some parts of the U.K., for example, where the population density is high, their murder rates are considerably lower than in low population density countries like the United States. You could argue that in countries where the population density is low, people were free to go wherever they wanted, they didn't want the government messing with them, and they didn't like a lot of rules and restrictions on their behavior, so they carried guns and were freer to use them." Thus in low-population-density countries or small families one does not have to learn conflict resolution as well. "That is the argument I make in my book *Chimpanzee Politics*," De Waal continues. "If you live in a community where you can successfully leave if there is conflict, then you avoid learning conflict resolution. But in crowded areas, if you move out, you just run into another group who doesn't want you. So in crowded places you need to learn to get along."

De Waal's theory closely parallels that of the Russian writer Pyotr Kropotkin, outlined in his book *Mutual Aid*, which was discussed in chapter 2 in the context of why people equate evolution with aggressive behavior and cutthroat competition. To reiterate, Darwin and Huxley, from ultracompetitive England, emphasized competition, whereas Kropotkin, from the more egalitarian Russia, pushed the cooperative side of nature. De Waal agrees that there is a linkage. "The Dutch are very consensus-building people. They live below the ocean, so they had to fight against it. If we live in a town where you are on the top of the heap and I am a peasant farmer, if the dam breaks you can't just sit in your castle. You've got to come down and join in helping to fix it. That makes for a more egalitarian society." Our evolved moral emotions can be tweaked by the environment? "We can really see that by comparing the Dutch political system with the British," De Waal notes, "The British monarchy is very isolated from the people, so the British have always had a very different attitude in terms of hierarchy, which is impossible in the Netherlands where even the Queen on occasion needs to show that she remains one of the people—for example, by riding her bicycle."

Another common myth about evolution is that it is based on "selfish genes," a metaphor employed by Richard Dawkins in his book of the same title. Many people take this to mean that evolution really is "red in tooth and claw." (As we'll see in chapter 11, Dawkins's book was a favorite of Jeffrey Skilling, the CEO of Enron.) Once again, De Waal is the evolutionary mythbuster. "A gene cannot be 'selfish,' of course, so this is just a metaphor," he begins. "The genome of humans has produced a very complex psychology that includes loyalty, solidarity, cooperation, pure altruism, and all that present in humans, chimpanzees, and other social animals." But genes do code for behaviors that will best serve the survival and reproduction of their owner, and it would then appear that evolution produces a selfish psychology. De Waal concedes this point, but notes that "there is also an unselfish psychology that in the long run has served us and these other social primates because they live in groups and they survive by mutual aid and cooperation."[3]

De Waal's research does support an evolutionary model for the origins of morality, so are we moral animals? Are chimps? "I would not call chimpanzees moral animals. But I do think they have many of the elements of moral behavior. The way I would put it is that our moral systems have made use of a much older psychology. Some moral philosophers might think that we have invented human morality, but that's not the case. Human morality is built on top of a psychology that can be seen in other primates."

Human economics is also built on top of a psychology that can be seen in other primates. In his 1982 book, *Chimpanzee Politics*, for example, De Waal describes behavior on the part of his chimp charges that are clearly "direct payment for services rendered," concluding from his collective observations: "Chimpanzee group life is like a market in power, sex, affection, support, intolerance and hostility. The two basic rules are 'one good turn deserves another' and 'an eye for an eye, a tooth for a tooth.'"[4] In his later book, *Peacemaking Among Primates*, De Waal presented substantive evidence that chimpanzees and other primates experience both sympathy and empathy and that they console one another after a fight, most notably in the form of the all-too-human hug or arm-about-the-shoulder embrace.[5]

Ape anecdotes also feature prominently in De Waal's work, most notably a remarkable incident that occurred on August 16, 1996, at Chicago's Brookfield Zoo, in which a three-year-old boy fell almost twenty feet down into the gorilla pen. The impact knocked him unconscious. An eight-year-old female gorilla named Binti Jua promptly scooped him up in her arms and cradled him in her lap, offering periodic reassuring pats on the back that he would be okay. "This behavior is simply an extension of what they would do within their own species," De Waal concludes, noting that such consoling behavior is commonly observed in his lab. "We measure consolation by simply waiting until a spontaneous fight occurs among our chimps, after which we note if bystanders approach the victim. Bystanders often embrace and groom distressed parties. It is not unusual for a climbing youngster to fall out of a tree and scream. It will immediately be surrounded by others who hold and cradle it."[6]

Conflict resolution is key to social relations, and De Waal is sure to mention that after a fight, bonobos famously employ a human favorite— make-up sex. Such conciliatory behavior is common in most social mammals, in fact. Dolphins and elephants, for example, show sympathy and empathy. "Elephants are known to use their trunk and tusks to lift up weak or fallen comrades. They also utter reassurance rumbles to distressed juveniles. Dolphins have been known to save companions by biting through harpoon lines, hauling them out of tuna nets in which they got entangled, and supporting sick companions close to the surface so as to keep them from drowning." There is one notable exception that all dog lovers will appreciate: cats. "Domestic cats are solitary hunters, so they don't need to maintain nice relationships with potential competitors. They do not live in the give-and-take society of primates, or social cats such as lions."[7]

Relationships are so critical in a social species that mammals, including and especially human and nonhuman primates, will learn to suppress aggression, resolve conflicts, and keep the peace to preserve friendships. This is the very foundation of trust and trade in an exchange economy. In studies with both chimpanzees and capuchin monkeys, for example, De Waal and his colleagues found that when two individuals work together on a task for which only one is rewarded with a desired food, if the reward recipient does not share that food with his task partner, the partner will refuse to participate in future tasks and expresses emotions that are clearly meant to convey displeasure at the injustice.[8] In another experiment in which two capuchin monkeys were trained to exchange a granite stone for a cucumber slice, they made the trade 95 percent of the time. But if one monkey received a grape instead—a delicacy capuchins greatly prefer over cucumbers (don't we all?)—the other monkey cooperated only 60 percent of the time, sometimes even refusing the cucumber slice altogether. In a third condition, in which one monkey received a grape without even having to swap a stone for it, the other monkey cooperated only 20 percent of the time, and in several instances became so outraged at the inequity of the outcome that it flung the cucumber slice back at the human experimenters![9] In a similar experiment with long-tailed macaques at the University of Zurich, primatologists Marina Cords and Sylvie Thurnheer found that when two monkeys had learned to cooperate with each other in a task that required both of their efforts in order to be rewarded with food, they were more likely to reconcile after a quarrel than those monkeys who had not learned to cooperate in the joint-effort task.[10]

Here we see the foundation of economic exchange in our primate cousins. Studying these mirrors of ourselves and our evolutionary past can tell us a great deal about what to expect from ourselves in various social situations, including and especially economic exchanges. Primate behavior is another "fossil," as it were, to piece together what life was like in our evolutionary environment, and how our responses to such pressures in the past are reflected in our behavior today. We evolved to care about exchange outcomes, most notably that they be fair. De Waal's colleague at Emory, Sarah Brosnan, studies inequity in nonhuman primates with an eye toward unraveling the evolution of negative responses to unequal social outcomes. "An aversion to inequity may promote beneficial cooperative interactions," Brosnan suggests, "because individuals who recognize that they are consistently getting less than a partner can look for another partner with whom to more successfully cooperate."[11] This would naturally lead to a selection for prosocial and cooperative behavior within groups—the very foundation of market exchange—and a selection for xenophobia and tribalism between groups. This is why extremely low ultimatum game offers are usually rejected—we are willing to forgo gain in order to prevent another from receiving an unjust reward. That is, we'll pay to punish fair-trade transgressors.

In fact, if we broaden our conception of an "economy" and what it means to trade, all primates trade in their social economy, using such currencies as food, grooming, conciliation, reciprocity, friendships, and alliances. De Waal has found in his research that chimps are more likely to share food with a chimp with whom they have groomed, and that capuchin monkeys tend to share food, grooming, and conciliation with fellow capuchin monkeys with whom they have traded such currencies before.[12] Similarly, the ethologist Nicola Koyama and her colleagues found a relationship between exchange of grooming and later social alliance in chimpanzees. It turns out that if chimp A groomed chimp B, then chimp B would be more likely to support chimp A in a fight with others the next day, especially if it was chimp A who was starting the fight. The scientists interpret this as meaning that chimpanzees curry political favor through trade in anticipation of a possible future need.[13] Primate politics indeed.

We therefore evolved the capacity to actually be moral animals, and not just give the appearance of being moral animals. That is, it is not enough to fake being a good person, you actually have to be a good person. Evolution explains how we got to be that way, and studying our closest living primates is one tool that helps us understand how that happened. Although De Waal admits that "human and animal altruism no doubt arose in the context of help to family members and help to those inclined to return the favor," he also points out that "to believe that a chimpanzee, or any other animal, helps another with the goal of getting help back in the future is to assume a planning capacity for which there is no evidence. We can be sure that most animals perform altruism without pay-offs in mind. In this sense, animal altruism is *more genuine* than most human altruism, since we do have an understanding of return benefits, hence are capable of mixed motivations that take self-interest into account."

Again, as I have argued throughout this book, these moral emotions are a proxy for a moral calculation done by evolution of traits selected for what is best for the individual or the group. The moral motivations and intentions that we attribute to human altruistic behavior are secondary and come after the fact. To employ another metaphor, we take an "intentional stance" (as the philosopher Daniel Dennett calls it[14]), in which we assume that other agents have intention, and that their intention may be good or evil. But this provokes a further evolutionary concept about the origination of moral behavior—that the evolved moral emotions are also proxies, where we not only assume that others of our group are really moral or immoral, but that we *are* moral or immoral, as well. In other words, the evolution of moral (or immoral) behaviors came first, the attribution of moral value came later. We really are moral primates.[15]

✦ ✦ ✦

"There's an old English proverb that says 'It is an equal failing to trust everyone and to trust no one.'" So begins Paul Zak, a professor of economics at Claremont Graduate University, who is taking his profession down to the molecular level in his search for the neurochemistry of trust, which he believes is grounded in oxytocin. "We know that trust is a very strong predictor of national prosperity, but I want to know what makes two people trust one another," he explains as we sit down in his Center for Neuroeconomics Studies, nestled in the bedroom community of Claremont, California.[16]

Zak is the oxytocin man. It says so right on his license plate. Tall and handsome with square shoulders and the physique of someone who works out regularly, Zak's firm grip and warm smile exude, well, trust. Trained in traditional economics, in the mid-1990s his research led him to connect trust to economic growth. A 1996 study on trust in forty-two countries, for example, asked people in their native language, "Generally speaking, would you say that most people can be trusted, or that you cannot be too careful in dealing with people?" The results were as diverse as they were striking. At the low end of the trust scale, only 3 percent of those surveyed in Brazil and 5 percent in Peru believe that their fellow citizens are trustworthy, compared to 65 percent of Norwegians and 60 percent of Swedes who trust one another. Falling in the middle of the scale were the United States at 36 percent and the United Kingdom at 44 percent. The rankings remain essentially unchanged even when they are controlled for income; trust is high in the countries of Scandinavia and East Asia but low in the countries of South America, Africa, and especially the former Communist bloc. "The simple correlation between national rates of investment (gross investment per Gross Domestic Product) and trust is strongly positive," Zak explains, so that "when trust is low, investment lags. The same positive correlation holds for GDP growth and trust."

Economic mechanics drive the relationship between trust and prosperity. Zak offers this explanation:

> Trust facilitates transactions by reducing the number of contingencies that must be considered when "doing a deal." A deal sealed with a handshake between principals can only occur in a high-trust situation. "Let the lawyers work out the details—we have a deal." Conversely, when trust is low, negotiations are protracted, and therefore more costly. When transaction costs are higher, fewer transactions occur and investment and economic growth are lower. Trust is among the most powerful stimulants for investment and economic growth that economists have discovered. In seeking to understand why some countries are poor and others are rich, it is, therefore, crucial to understand the foundation for interpersonal trust.

Perhaps this is why countries that have higher rates of generalized trust show higher rates of return on national stock markets.[17] From these and other studies, Zak concluded that in order for a nation to achieve prosperity it is vital to maximize positive social interactions among its members in order to increase trust.

The list of positive social interactions identified by Zak's research will surprise no one living in a liberal democracy with relatively free markets: *protection of civil liberties*, *freedom of the press*, *freedom of association*, *freedom of travel* (good roads and reliable infrastructure), *freedom of communication* (working phone systems), *mass education*, *a reliable banking system*, *a sound currency*, and especially *the freedom to trade*.[18] He even found a connection between a clean environment and trust, whereby people in countries with polluted environments show higher levels of estrogen antagonists, which lowers their levels of oxytocin—and thus their feelings of trust. Zak went so far as to compute the differences in living standards that trust can effect, whereby "a 15 percent increase in the proportion of people in a country who think others are trustworthy raises income per person by 1 percent per year for every year thereafter." For example, increasing levels of trust in the United States from its present 36 percent to 51 percent would raise the average income for every man, woman, and child in the country by $400 per year, or $30,000 over a lifetime.[19] Trust has fiscal benefits.

The connection between social interactions and trust can also be seen in the laboratory when subjects participate in experiments utilizing the Prisoner's Dilemma. In the scenario, each of two subjects pretends to be a prisoner arrested for a crime. Each is independently made the same offer, and both assume the other has been presented the same deal. Imagine that you are the subject and these are your options: (1) If you and your partner both cooperate, then you each get one year in jail; (2) If you confess that both you and your partner committed the crime, then you go free and your partner gets three years; (3) If your partner confesses and you don't, then you receive the three-year penalty while he goes free; (4) If you both confess, then you each get two years. If you defect on your partner and confess, then you will get either zero or two years, depending on what he does. If you cooperate and stay quiet, you get either one or three years, again depending on his response. The logical choice then is to defect. Of course, your partner is likely going to make the same calculation as you, which means he too will defect, guaranteeing that you will receive a two-year punishment. However, you also then come to the conclusion that he is probably computing this same strategy, and hope that as a result he will realize that you should both cooperate. But then, if that is his line of reasoning, perhaps he'll defect in hopes that you cooperate assuming that he will cooperate, getting him off the hook while you get hit with the three-year penalty. This is why it is called a dilemma.

When the game is played just once, both people typically defect. But even when the game is played over a fixed number of rounds, defection is the norm, because each person realizes that the other player will defect on the last round, and thus is spurred to defect in the second to last round, which spurs the other player to defect a round earlier, and so forth until both players decide to defect from the first round. However, when the game is played as something more similar to the real world, with the game played over an unknown number of rounds—as in a negotiation process where the number of steps are typically unknown—both players keep track of what the other has been doing round by round and cooperation prevails. Over and over it has been shown that the best strategy is "tit for tat" with no initial defection. That is, the most self-benefiting thing to do in the long run is to begin by trusting and cooperating and then do whatever your partner does. In such a Prisoner's Dilemma contest, the winning strategy, called "Firm but Fair," dictates cooperation with cooperators, cooperation after a mutual defection, not playing with constant defectors, and defecting with partners who always cooperate (otherwise known as suckers).[20] Even more realistic simulations include the "Many Person Dilemma," in which one player interacts with several other players, and in conditions where subjects are allowed to accumulate experience with the other players, thereby giving them an opportunity to establish trust. Finally, the Prisoner's Dilemma protocol has been applied to the real world of business negotiations, marital disputes, and cold war strategies. It turns out that in both computer simulations and the real world, cooperation is by far the dominant strategy.[21]

In a related experiment on cooperation, nine subjects were each given $5. If five or more of the nine cooperated by donating their $5 to a general pot, all nine would receive $10. Although it pays to be a cooperator (you get $10 instead of $5), it pays even more to be a defector ($15 instead of $5), as long as at least five other people cooperate. The results were mixed, with many groups of nine subjects failing to achieve the critical mass of five cooperators, because there was no cooperation. Then the experimenters added a step: members of some groups were given the opportunity to discuss their strategy options before playing. Those groups that interacted before playing averaged eight cooperators, and 100 percent of these groups earned cooperative bonuses. By sharp contrast, those groups that did not interact before playing earned bonuses only 60 percent of the time.[22] In a similar experiment on social dilemmas, psychologist Robyn Dawes found that groups given the opportunity to communicate face-to-face were more likely to cooperate than those who were not. "It is not just the successful group that prevails," Dawes concluded, "but the individuals who have a propensity to form such groups."[23]

Where in the brain are these dilemmas resolved? Employing the Prisoner's Dilemma protocol while scanning thirty-six subjects' brains using an fMRI, James Rilling and his colleagues at Emory University found that in cooperators the brain areas that lit up were the same regions activated in response to such stimuli as desserts, money, cocaine, and beautiful faces. Specifically, the neurons most responsive were those rich in dopamine located in the *anteroventral striatum* in the middle of the brain—the so-called pleasure center. Tellingly, cooperative subjects reported increased feelings of trust toward and camaraderie with likeminded partners.[24]

These findings make sense in an evolutionary model, where informal and noncodified rules of conduct developed within small bands of hunter-gatherers because knowing all the other players in the game leads to the evolution of cooperation.[25] The psychological impulse to form relationships and alliances is the deeper cause that lies beneath the moral sentiment of trust, and trade is an effective medium that allows people to create trusting relationships with and form attachments to other trustworthy people. And to reinforce my thesis one more time, it is not enough to fake being a cooperator, because over time and with experience, deceivers are usually flushed out. You actually have to believe you are a cooperator, and there is no surer way to believe you are a cooperator than to actually be one, believe it, and mean it. As Yogi Berra counseled, "Always go to other people's funerals; otherwise, they won't go to yours."

From behavior to brains to blood, the molecular basis of trust begins at the most fundamental level of human relations. In her book *Why We Love*, anthropologist Helen Fisher makes a distinction between lust and love. Lust, she says, is enhanced by dopamine, the neurohormone produced by the hypothalamus that we have already seen is directly associated with reward and pleasure. Fisher shows that it also triggers the release of testosterone, the hormone that drives sexual desire, and thus dopamine is implicated in yet another vital component of human relations—the yearning desire for another.[26] Love, by contrast, is the emotion of attachment reinforced by oxytocin, a hormone synthesized in the hypothalamus and secreted into the blood by the pituitary. In women, oxytocin stimulates birth contractions, lactation, and maternal bonding with a nursing infant. In both women and men it increases during sex and surges at orgasm, playing a role in pair bonding, an evolutionary adaptation for the long-term care of helpless infants. Monogamous species, for example, secrete more oxytocin during sex than polygamous species.[27]

Here we return to the connection between trust, trade, and oxytocin, and Paul Zak's theory that there is a direct connection between the three. "Oxytocin and testosterone are two branches of this trust-distrust system, and as we go through the world we are constantly balancing these levels of trust and distrust," Zak explained as he recounted some of the experiments he and his team have been running in the center. In one experiment, for example, Zak found that oxytocin increases "when a person observes that someone wants to trust him or her." It turns out, in fact, that with the exception of the 2 percent of the population who are sociopathic, when someone trusts, it triggers the release of oxytocin. Although this is the case for both men and women, Zak has found, "When women get a boost of oxytocin they are more likely to reciprocate than men are. Men are more sensitive to violations of social norms, and when the norms are broken men get a disproportional rise in testosterone. Women don't like being distrusted, but they don't have this heated and angry response that men get, which is related to testosterone."

Oxytocin is deeply involved in the attachment process, whereas testosterone is intensely caught up in the enforcement of the social norms (which may help to explain why far more men go into such professions as the military and law enforcement). In exchange games, the more subjects are behaving in trusting ways, the more money they exchange and the higher the levels of oxytocin that are released by the brain.[28] When Zak asks them *why* are they giving up so much money, the subjects report that it "just feels right." The moral emotion drives behavior, even if the moral calculation beneath the emotion is invisible.

Skeptics might reasonably ask whether oxytocin is the cause or the effect of trust. "To control for that," Zak says, "we set up an experimental condition where instead of subjects freely choosing to trust someone, we had them randomly pull out a Ping-Pong ball from a box that would determine how much money was given or received, and in those cases there was no significant change in oxytocin levels." To find out if cooperation and trust lead to the release of oxytocin or if increased levels of oxytocin lead to more cooperation and trust, Zak infused oxytocin into subjects' brains through a nose spray containing oxytocin, which is quickly absorbed by the body; he discovered that it causes them to act more cooperatively.

Zak has adduced a number of additional lines of evidence to support his thesis. *Trust and happiness*: People who trust and are trustworthy report being happier. *Trust and touch*: We all know how good it feels to be touched by someone, so Zak ran an experiment in which he gave subjects in an exchange game a massage by a licensed massage therapist, which led those who received the massage and a signal of trust to offer up to 250 percent more than subjects who did not receive them. *Trust and smell*: Oxytocin may also be mediated by smell, Zak suggests, noting that there are oxytocin receptors on the olfactory bulb and citing an experiment in which the smell of a mother's own newborn baby triggers the release of oxytocin and powerful feelings of attachment. *Trust and neglect*: Animals that are abused or neglected shortly after birth show a loss of regions of the brain that have oxytocin receptors, and those animals become withdrawn, socially inappropriate, and depressed.[29] The implications of Zak's research are profound. "Oxytocin is the social glue of society. It is what keeps us together as a civilization. If we didn't have something in our head that indicated who we should trust and who we should not, civilization wouldn't work. We're social animals and we need a trust detector in our heads."

Zak's new findings about oxytocin get at something very deep in the evolutionary origins of morality: the role of evil people in society. In my earlier book, *The Science of Good and Evil*, I attributed evil to our dual dispositional nature and the fact that in addition to being trusting, cooperative, and altruistic, we are also distrusting, competitive, and selfish, and that the evolutionary forces that led us to be prosocial with our fellow in-group members also led us to be tribal and xenophobic against out-group strangers. Zak pushes the evolutionary model further by looking not just at the potential for evil that resides in all of us, but at the anomaly of evil that lurks within 2 percent of individuals, those who are sociopaths. While the research shows that sociopaths comprise 3 to 4 percent of the male population and less than 1 percent of the female population, they are believed to account for 20 percent of the United States prison population and between 33 percent and 80 percent of the chronic criminal offender population. Altogether, these individuals may account for half of all crimes in the United States.[30]

In his social game experiments, Zak typically finds that about 2 percent of his subjects (he calls them "bastards") do not respond to oxytocin or other social cues that normally encourage trust and cooperation. "These individuals have a dysfunction in their oxytocin release and have an identifiable difficulty attaching to others." But instead of seeing this as nothing more than a biological accident in the wiring of the brain, Zak thinks there may be adaptive evolutionary reasons behind such misfirings. "Bastards are necessary from an evolutionary standpoint because they keep the physiologic balance between appropriate levels of trust and distrust optimally tuned. Without these exceedingly selfish people, humans might have evolved into being unconditional trusters. If so, we would become susceptible to invasion by those who would prey on our perfectly trusting nature." Knowing my interest in and many trips to the Galápagos, Zak reminded me of the iguanas and other animals there that evolved without human (or other) predators and are therefore largely unafraid of us. You can walk right up to them and take photographs from inches away, for example, and in previous centuries, before an environmental ethic had developed, sailors helped themselves to free food on the hoof whenever they wanted. "The two percent of folks who appear to unconditionally prey on others keep the rest of us on our toes," he concludes, "both individually and as a species."

As part of his evidence, Zak cites the case of a woman who has a rare genetic disorder that caused her amygdala to calcify and die. "The amygdala is a primary target for oxytocin in the brain that helps maintain the trust/distrust balance," he notes, "and she has difficulty reading the trustworthiness of faces, is very impulsive in her decision-making, and is terrible with money. She has normal I.Q., but is often a target for unscrupulous predators who sense her unconditional trust in others and take money from her." Caltech's Ralph Adolphs examined the woman for other deficiencies and found that she is unable to recognize fear in the face of others. When you look at another person's face, your eyes rapidly dart about scanning for clues and collecting data about its emotional expressions, which, according to psychologist Paul Ekman, are evolved to be universal and thus identifiable. The amygdala-damaged woman, by contrast, stares straight at the face without scanning for details, and thus is unable to make an emotional assessment.[31] During social interactions, most of us maintain a balance between trust (mediated by oxytocin) and fear (mediated by epinephrine, norepinephrine, and other stress hormones), and we quickly and unconsciously adapt to our environment and the people in it. But when that balance is broken, so too is our trust detector.[32]

To this example, Zak adds the extensive research on our emotional responses to violations of social norms of cooperation. "We can all be pushed into the enforcement role when we are treated badly," he notes, reminding me how we have all chased down a car that cut us off in traffic just so that we could express our displeasure with the appropriate hand signal. "Why would you waste your time doing this? Because that jerk needs to learn that you can't treat people like that!" Zak laughs, knowing what I know about this example—it's more of a guy thing to do. "Studies by my lab show that physiologically men are much more sensitive than women to violations of social norms. Indeed, in experiments we have run, the more a man is distrusted (using the trust game), the greater his rise in the high-octane version of testosterone (called dihydrotestosterone)." Women dislike social norm violations as much as men do, of course, but they do not have the same heated response to it that leads so many men, especially young high-testosterone men, to respond with violence. Indeed, men even seem to get a burst of pleasure—or at least, reward activation—when they punish a norm violator, as brain scans have shown high activity in the NAcc reward center (discussed in chapter 6), which is fueled by dopamine.

✦ ✦ ✦

Zak's model fits well with my own, that we evolved an innate sense of right and wrong that is expressed through the moral emotions, and that free trade is an integral component to breaking down the normal tribal barriers blocking trust. That being the case, economic transactions can occur with a minimum of top-down interference. A *shadow of enforcement*—a hint of potential punishment for a norm violation—is all that is needed for most people in most circumstances to grease the wheels of commerce. A law enforcement dummy placed along a stretch of highway, for instance, has been shown to act as a shadow of enforcement to get motorists to slow down not because they thought it was an actual enforcer but because it reminded them of the law.

Trade makes people more trusting and trustworthy, which makes them more inclined to trade, which increases trust . . . creating a self-enforcing cycle of trust, trade, freedom, and prosperity. Each year, the Heritage Foundation publishes a report on economic freedom from around the world, and it is instructive to see the correlation between its ratings and Paul Zak's data on national trust within countries:

1. Correlation between economic freedom and trust = 0.31
2. Correlation between economic freedom and per capita GDP = 0.74
3. Correlation between trust and per capita GDP = 0.46

These figures indicate that the relationship between economic freedom and trust is high, but disappears when per capita GDP is included as a control, and hence the same environments that produce high trust also produce high economic freedom. Remember too that an increase in per capita GDP raises trust, and an increase in trust raises growth in per capita GDP, so again we find a positive feedback loop driving the system.

An additional evolutionary connection between trust and trade can be seen in a two-part experiment on cooperation and cheating conducted by Dan Chiappe and his colleagues at California State University, Long Beach, in which subjects first categorized individuals as cheaters, cooperators, or neither, based on a written description of how they behaved in an exchange involving, for example, borrowing and paying back (or not) money. After reading the description, each subject was asked to rate how important it was to remember the individuals, using a seven-point scale. In the second experiment, participants categorized the individuals on the seven-point scale as before, but then had the opportunity to look at photographs of their faces. They were then presented with a face recognition test. The first experiment found that cheaters were rated more important to remember than cooperators—especially when a greater amount of resources was involved—and cooperators were rated more important to remember than those categorized as neutral. The second experiment showed that cheaters were looked at longer and the subjects had a better memory for their faces.[33]

Why should this be? Because from an evolutionary economics perspective, cheaters—like the 2 percent who routinely break society's laws—keep cooperators on their moral toes, so it is vital to be able to discriminate between cheaters and cooperators. As Chiappe explained, "Everything else being equal, knowing that a person cooperated may not tell us much about their character. Knowing that they cheated would be more relevant. This is because cheaters have to give the appearance of being trustworthy and thus they may have to cooperate much of the time." Indeed, most people most of the time in most circumstances are cooperative, so it takes a lot more mental data storage space to focus on and remember details about them (such as their faces) compared to the small handful of cheaters, whose social norm violations stand out as the exception and thus must be recalled for future transactions. And, as research in memory has consistently demonstrated, we are more likely to recall unusual events than we are common occurrences. The findings also indicate that we remember cooperators more than neutral people.

This too makes sense from an evolutionary perspective on the importance of social contracts since the memory gradient from cheaters to cooperators to neutrals would be adaptive in our management of social relations in all forms of human exchange, especially economic trade.

If this explanation is true, there should be neural networks associated with cooperation in an exchange task, and sure enough, the George Mason University neuroeconomist Kevin McCabe and his colleagues, in an fMRI study of subjects who participated in a "trust and reciprocity" game with either another person or a computer, found that players who were more trusting and cooperative showed more brain activity in Brodmann Area 10, associated with reading others' intentions, and more activity in the limbic system, associated with the processing of emotions. Tellingly, such activity was not seen in this brain region of players who either played against computers or engaged in noncooperative games, which makes perfect sense in a social primate species that evolved moral emotions that drive such prosocial behaviors.[34]

◆ ◆ ◆

Evolutionary economics suggests that instead of thinking of human culture as being cleaved by different national economic institutions and varying individualized personal values, we should think of it as an evolved human nature that gives rise to a set of core institutions and values that vary in details across different cultures. In this sense, *Homo economicus* lives, but in a mutated form: *Homo evonomicus*.

As a test of this interpretation, over the past quarter century hundreds of experiments in behavioral economics have been conducted on people from dozens of countries around the world, including fifteen small-scale indigenous tribes. Take the Ultimatum Game, where one player proposes how to divide a sum of money and the second player accepts or rejects the proposal. While American subjects—a relatively homogeneous population—typically propose (and accept) splits along the lines of 70-30, the average offer by people from these small-scale tribes varies systematically, from a minimum of 26 percent for the Machiguenga tribe in Peru to a maximum of 58 percent for the Lamelara tribe in Indonesia. But the variation is correlated with the prevailing occupation of the people, where those whose economies are large and more market-integrated tend to make higher offers than those whose economies are at a more subsistence and less market-integrated level.[35]

In a comprehensive overview of the substantial database now accumulated on how humans from around the world and in a variety of cultures behave economically, behavioral economist Colin Camerer and his coauthors conclude that people everywhere have a set of consistent social preferences. These include "subjects care about fairness and reciprocity, are willing to change the distribution of material outcomes among others at a personal cost to themselves, and reward those who act in a pro-social manner while punishing those who do not, even when these actions are costly."[36] Furthermore, even though human cultures vary enormously, with vastly different forms of social organization and institutions, kinship systems, and environmental conditions, there nevertheless remains a set of core features of human economic nature that clearly have an evolutionary basis to them. To wit, in summing up the cross-cultural data, the authors of the summary study draw five general conclusions:

1. The selfishness axiom is not supported in any society studied.
2. There is considerably more behavioral variability across groups than had been found in previous research.
3. Group-level differences in economic organization and the degree of market integration explain a substantial portion of the behavioral variation across societies: the higher the degree of market integration and the higher the payoffs to cooperation, the greater the level of prosociality found in experimental games.
4. Individual-level economic and demographic variables do not explain behavior either within or across groups.
5. Behavior in the experiments is generally consistent with economic patterns of everyday life in these societies.[37]

In rejecting *Homo economicus*, however, Camerer and colleagues also rebuff such suggested alternatives as *Homo altruisticus* or *Homo reciprocans* as being equally oversimplified: "The diversity of behaviors we have observed leads us to doubt the wisdom of this approach." My proposed *Homo evonomicus*, however, incorporates the dual dispositional nature of humans by acknowledging that we are both selfish and altruistic, while at the same time recognizing that this evolved nature is sensitive to tweaking by cultures and institutions.[38]

✦ ✦ ✦

Any theory of economics must begin with a sound theory of human nature. Evolutionary economics redefines the borders of our nature, showing just how driven we are by ancient programs designed for another time and another place. But we also evolved the *adaptation of adaptability*, and it is here that we see how and why humans behave as we do in such social institutions as markets. We cooperate for the same reason we copulate—because it feels good. On a deeper evolutionary level, the reason cooperating feels good is because it is good for us, individually and as a species. Trust and cooperation lead to a viable free market of exchange, and free markets lead to greater trust and cooperation—the very model of a complex adaptive system that learns as it develops.

Over the past six million years of primate evolution, as our brains increased in size, particularly the frontal cortex, we gained the ability to better control our impulsive emotions and delay gratification, as well as form complex social coalitions. Between-group competition for limited resources led to selection for competitiveness and selfishness, but at the same time it led to within-group cooperation and selflessness that enhanced the fitness level of individual members of the group and the group itself. This evolutionary interpretation bodes well for our species. Although recent history is not encouraging, the long-term trend over the past half millennium has been toward greater inclusiveness and more liberties for more people in more places.

10

THE SCIENCE OF GOOD RULES

In 1982, two other men and I founded the three-thousand-mile nonstop transcontinental bicycle Race Across America (RAAM) from Los Angeles to New York, sponsored by Budweiser and televised on ABC's *Wide World of Sports*. The rules were simple: each cyclist takes the same route, with a support vehicle and crew that follows behind providing food, drink, and equipment, and no drafting behind or hanging on to a vehicle allowed. The race started on the Santa Monica Pier in California. The first cyclist to reach the Empire State Building in New York City would be declared the winner. That was the entire set of rules, which we didn't even bother to write down.

All four of us finished the race.[1] At the time, I figured that this would be a one-off event, but ABC won an Emmy for sports programming for their coverage of the race, which drew considerable interest from cyclists and endurance athletes from other sports. Ultramarathon events were generally gaining popularity, and dozens of cyclists wanted to compete the following year, so we set a date for the next race, held a qualifying event, and began to outline some rules.

During that first race, for example, it wasn't clear what to do with a rider who went off course—could he get a ride in his support vehicle back to the route where he left it? (Yes.) Could he be driven up the route the same distance he rode off course? (No.) Although no drafting was allowed, it can often be windy out on the open plains, and if a crosswind blew from your left when your support vehicle came alongside to hand off water bottles and food, you got a noticeable drafting effect. Not to mention that ten days was a long time to ride by yourself, so it was psychologically advantageous to have your support crew to talk to for long stretches. So we had to draft extensive rules defining how long a handoff could last (one minute), how many times an hour (four), and build in exceptions to the rule, such as if the temperature exceeded 100 degrees, in which case the number of handoffs was unrestricted.

Once you start writing down what people can and cannot do, the list grows exponentially. As the years moved on and the number of racers grew, the rulebook expanded with it. Women entered in 1984, so we added rules about gender divisions. Cyclists over fifty and sixty years of age wanted to race, so we added rules about age divisions. Four-man relay teams entered in 1989, so we created a new set of rules just for them, which subsequently had to be expanded to encompass two-person relay teams, men-and-women relay teams, age division relay teams, and even corporate relay teams. Every year something would happen that led to more rules. In the 1989 race, one of the competitors was riding slowly up the long grade of Oak Creek Canyon from Sedona to Flagstaff, Arizona, with his two vans and motorhome all caravanning behind him, preventing cars from safely passing. This went on for miles until someone called the police, but by the time the officers arrived they could only find the next rider back, whom they stopped on the side of the road, thereby disrupting his pace and costing him time, which we had to subtract from his overall finishing time. Another year, also in Arizona, we had a similar problem on a busy stretch of Highway 89, after which someone called the Arizona Department of Transportation to complain, resulting in a postrace ruling by the DOT that RAAM could pass through Arizona only with no following vehicle and only during the day. As this would have obviated the nonstop nature of the race, I had to negotiate a deal with the Arizona DOT with a proviso in the rules that read: "The Follow Vehicle may not impede following traffic for more than one minute. The Follow Vehicle must pull off the road and let traffic pass when five or more vehicles are waiting to pass regardless of time. During the day the rider may proceed alone, with the Follow Vehicle catching up once traffic is clear. At night the rider must also pull off the road."

During an especially hot year, one of the riders happened upon a small hotel pool while passing through a diminutive Western town, so he dismounted his bike and leaped into the pool, fully attired in cycling clothes, shoes, gloves, and helmet. Someone called the police, and once again by the time they arrived the pool perpetrator was gone, so they pulled over the next competitor who happened along. This led to yet another rule, this one prohibiting competitors from swimming in public pools without permission from the owner. Most of the racers enjoyed listening to music, either from small earbud phones or from speakers mounted on the roof of their following vehicle, which led to two additional problems: one, blasting music through tiny earphones jammed in your ears makes it difficult to hear oncoming traffic, ambulances, and the like; and two, passing through small towns in the middle of the night blasting rock 'n' roll music from loudspeakers tends to awaken the locals. Thus, two new rules were added, one restricting cyclists to only one earphone and the other curtailing the broadcasting of music during hours of darkness.

All of these new rules, of course, required the addition of appropriate punishments for violations, which we assessed in time penalties. However, an exhausted cyclist who is forced by an official to take time off the bike in the middle of the race may actually benefit from the penalty, so we had to write even more rules about where and when the penalties would be served, which we determined would be in a penalty "box" ten miles from the finish line (and, yes, we had cyclists passed by competitors while sitting there on the side of the road). More rules and penalties meant more officials needed to assess them, and therefore more potential for subjective misjudgments on the part of officials (who were often sleep-deprived and exhausted themselves), so we had to add yet another set of rules that allowed the cyclists to challenge the officials' assessed penalties, and a set of guidelines for the race director to make a final evaluation before the end of the race if such a challenge was made, as well as a postrace board to hear one final appeal by the athlete if he or she felt that both the official and the race director made the wrong decision. Finally, we created a nonprofit governing body—the Ultra-Marathon Cycling Association (UMCA)—to oversee the entire sport, especially the development and adjudication of the rules.[2]

The structure and development of this sporting event and the rules that govern it—as quotidian an example as it is—serve as an analogue for society at large. In its simplicity, sports can help clarify and illuminate the evolution and operation of more complex and nuanced social institutions, just as good rules make good competitors, good walls make good neighbors, and good laws make good citizens. People naturally want what is best for themselves, but most people also want what is fair. Without a structure in place to create and enforce firm and fair rules to meet both of these needs, people become more self-centered than other-centered, and if you go far enough down that path, it's *bellum omnium contra omnes*, "the war of all against all."[3]

◆   ◆   ◆

If governing bodies make the rules for sporting events, who makes the rules for society? Institutions. Institutions generate both the informal social norms and the formal legal rules that govern individual behavior and structure interactions between individuals in social situations.[4] The Washington University economist Douglass North, who nabbed a Nobel Prize for his pioneering research on the relationships between institutions and economics (a field now called *institutional economics*), outlined these two general types of institutions. *Informal institutions* are structures such as traditions, customs, and values. These are bottom-up, self-organized institutions that evolve slowly and change when nudged by cultural trends and shifting traditions. *Formal institutions* are rules and laws that are intentionally designed from the top down, such as constitutions and legal codes, and for which change typically comes about only disruptively through new legislation or, when legal means are ineffective, through wars and revolutions.[5] "Institutions are the humanly devised constraints that structure human interaction," North explained in his Nobel Prize lecture. "They are made up of formal constraints (rules, laws, constitutions), informal constraints (norms of behavior, conventions, and self-imposed codes of conduct), and their enforcement characteristics. Together they define the incentive structure of societies and specifically economies." In keeping with the sports analogue, North notes, "If institutions are the rules of the game, organizations and their entrepreneurs are the players." Organizations, for example, are bound by a common goal and include "political bodies (political parties, the Senate, a city council, regulatory bodies), economic bodies (firms, trade unions, family farms, cooperatives), social bodies (churches, clubs, athletic associations), educational bodies (schools, universities, vocational training centers)."[6]

According to the Vanderbilt University law professor Erin Ann O'Hara, informal and formal institutions can work hand in hand. As an example, she points to two institutions that make trade possible: Informal *reliable middlemen* broker deals between strangers, each of whom knows and trusts the middleman, even if the trading partners do not know each other. Retailers whom you know in your local community serve as middlemen between manufacturers (which may be on the other side of the planet) and you; eBay is a cyberspace middleman, with a system of seller and customer evaluation in place to monitor who is trustworthy and who is not. The reputation of the middleman, then, is crucial to both parties. In tandem, formal *contract law* enforces agreements between strangers that could not otherwise be adjudicated in a dispute. O'Hara notes that contract law creates a sense of trust in three ways: (1) If both parties know that their contract will be enforced through the authority of the courts, they can assume a trusting position in a transaction with a stranger even without a middleman; (2) Flexibility of contracts allows parties to tailor their agreements to their specific wishes; and (3) Rules of contract interpretation and party conduct set precedents for future contracts, as well as courts, who must interpret them when there are charges of breach of contract. "Contract doctrines work to prevent parties from behaving in opportunistic fashion with one another," O'Hara explains, "while at the same time providing contracting parties with enormous flexibility to structure their transactions and relationships in order to address their individualized trust concerns."[7]

In addition, economic history is littered with cases that demonstrate our transition from informal to formal institutions as we have made the transition from personal to impersonal exchange; that is, from trade between family and friends in small communities to trade between strangers and foreigners in large societies.[8] In the Wild West of the American frontier, before the long arm of the law could reach that far from Washington, D.C., property rights were established and enforced through such informal institutions as land clubs, mining districts, and cattlemen's associations. Squatters who believed that possession was nine-tenths of the law set up land clubs to justify their settlements and enforce ownership of their newly acquired land before the government did it more formally with the Homestead Act. Miners did something similar by setting up mining districts and clubs to run them, establishing informal rules that would later evolve into public mining law. Cattlemen branded the hindquarters of their calves as a literal stamp of ownership, a rule enforced through informal cattle associations that endorsed the employment of hired gunmen as a prelude to a whole suite of laws later passed to do the same thing formally.[9]

A more recent example of such a transition from informal to formal institutions can be seen in the shift from "open range" laws that governed cattle ranching in Shasta County, California, to "closed range" ordinances in subregions of the county starting in 1945. In the open range system, ranchers were not legally liable for damages when their cattle wandered across unfenced land onto other ranchers' property. In 1945, a new state law authorized a closed range ordinance that made ranchers liable for damages to other ranchers caused by their livestock. And yet, a study by the economist Robert Ellikson shows that before the law, county ranchers voluntarily chose to report and compensate for damage despite what economic and legal theory predicted. Based on the informal institutions set up under the open range system by the ranchers, "neighbors in fact are strongly inclined to cooperate, but they achieve cooperative outcomes not by bargaining from legally established entitlements, as the parable supposes, but rather developing and enforcing adaptive norms of neighborliness that trump formal legal arrangements." And like most bottom-up self-organized systems, the informal ranching rules trumped the formal legal laws established by outsiders, resulting in "coordination to mutual advantage without supervision by the state."[10] The informal institution in this case worked better than the formal because the ranchers were intimately knowledgeable about their livestock, land, and business, whereas the judges, attorneys, and insurance adjusters were not. Under the informal system, neighborly rules developed whereby ranchers would notify each other when an animal strayed; if there were damages, they were resolved through barter and reciprocity; and violators of the informal system were dealt with through gossip and shunning.

A similar study by the economist Ronald Coase demonstrated that it is not necessary for the government to install formal legal controls over "public good" in order to solve the problem of free riding (where people who do not pay for the services benefit nonetheless and at the higher expense of those who do pay for it). In the case of privately held lighthouses, the public good problem was a chimera. Coase showed that lighthouse owners contracted with port authorities to collect fees from arriving ships at portside before loading or unloading their cargo, thereby solving the free-riding problem through bottom-up informal rules rather than top-down formal laws.[11]

The rub is in finding the balance between informal and formal institutions. The former are fluid and flexible and easy to change; the latter are not. Once formal institutions are established they are difficult to disestablish. They are sticky. They build up momentum. They become self-perpetuating. Discretionary monies used to set up a formal institution soon become set in budgetary stone. People's jobs are at stake. Livelihoods are on the line. What was once superfluous becomes indispensable. Clearly we need some formal institutions, but we must be prudent in our judgments about when, where, and how often we need to push informal rules into formal laws.

✦ ✦ ✦

I was thinking about such questions one day when I was driving my daughter home from school and we decided to order take-out dinner from a Thai restaurant in Pasadena. When we arrived at the restaurant I swapped some slips of paper for the food and we took our dinner home and consumed it without a second thought.

Giving a second thought to what is behind such a mundane activity reveals the elaborate networks of trust that we all take for granted. In the late afternoon of that particular day, I left my home to pick up my daughter, confident that the local police would guard the house against burglars, trespassers, and squatters, that the fire department would protect it from being consumed by flames, that my mortgage company would not take it away from me without due process, and that the military would protect it from foreign invaders. Pulling out of my driveway, I was assured that my automobile manufacturer had met a whole suite of quality control and safety standards—the wheels would not fall off, the accelerator would not jam, the brakes would consistently and without fail decelerate the vehicle and bring it to a full stop when pressed, and the three-point belt would hold me fast to the seat in the event of an impact, which would also reliably trigger the air bags. The roads upon which I drove that day had met a different set of quality control and safety standards, reassuring me that sinkholes would not suddenly appear before me, that traffic lights would properly engage to prevent intersection accidents, that lane widths would accommodate all vehicle sizes, that neither extreme heat nor cold would distort the shape of the road, that rainwater would be properly drained off the pavement, and that maintenance crews would routinely repair potholes, repaint lane lines, and uphold the quality standards through even the most severe weather conditions.

Arriving at my daughter's school, I didn't even notice all the precautions taken to protect the safety and well-being of the children—fences and locking gates, a crossing guard, a gate entrance monitor, and detailed rules about the pickup sequence that ensured the process would be safe and efficient. And, of course, there is the complex of conditions during the school day that I neither see nor even think about—the school's accreditation is maintained, the curriculum has been vetted and approved by a state curriculum board, the teachers are trained and certified, the classrooms are clean and well supplied, the gymnasium showers are regularly cleaned and sanitized, the cafeteria food is safe to eat, the lawns are mowed and grounds kept up, and the buildings repaired when needed. In fact, the only time any of us parents ever bothers even to think of such matters is when there is a problem with one of them, which is so rare that when it does happen it becomes a matter of much gossip.

On the way to the restaurant we decided to phone in our order, which triggered another net of social and technological conditions that made it possible for my cell phone signal to reach the information operator, who then transferred me to the restaurant, where the hostess took our order and asked for my name and phone number, thereby increasing the level of trust that I would actually show up and pay for the food. Arriving at the restaurant, I trusted that they would accept my slips of paper in exchange for the food because of the countless experiences we have all had in which the currency of our society is sound and its value stable. And when it comes to money, the network of trust between individuals, businesses, banks, moneylenders, and numerous other financial institutions—including and especially the government itself—is so convoluted that most of us have only an inkling how it all stays together. But that's the beauty of the trust web—we don't need to know how it works, just that it does. We only notice it when it breaks.

As we were waiting for our food, I noticed that the restaurant's new owners had removed the previously low-hanging ceiling in order to provide more vertical space, painting the supporting beams, ventilation tubes, and electrical pipes flat black. What I did not notice was the integrity of the building itself, which I just assumed to be water resistant, structurally sound, and—this being Southern California—relatively earthquake proof. The hostess offered us a glass of ice water while we waited, which we imbibed without wondering if it was potable, just presuming that the water supplier for the city of Pasadena had met the state's standards for water purity. Picking up our food, I never thought for a moment that I should first have it tested for poison or botulism, assuming that the local restaurant inspectors who gave this establishment an "A" rating were not being paid off by the owner and that they have a list of standards established by sanitation experts that the inspector ran through and approved. And based on past experience I also trusted that the Zagat rating indicating that this restaurant serves the best Thai food in Los Angeles was reliable.

The networks of trust and the social institutions that enforce them are so deeply embedded in our cultures and psyches that they are virtually invisible. Like the bird that is unaware of the air through which it flies, or the fish that does not notice the water in which it swims, the web of trust in which we live is so all-encompassing and so deeply entrenched in everything we do that we do not even know it is there until something goes wrong. Even when a network breaks, additional layers of trust are in place. If the police or fire department fails to protect my home, I have confidence that the insurance company with whom I contracted will honor our agreement to indemnify me against loss. Multiple levels of financial security now guarantee that a run on my bank will not cause it to go under, that the currency will remain relatively stable, that runaway inflation will be kept in check, and that my long-term investments represent something more tangible than binary digits on a computer screen. A powerful legal system affords me redress against the automobile manufacturer for negligence in its design, against the city for neglecting the roads, against the restaurant for failing to deliver edible food, against the state for providing contaminated water, and even against the government for not enforcing the standards it set or for establishing inadequate standards in the first place.

The web of trust is cast upon everything we do, and without it we would be like strangers in a strange land.

◆ ◆ ◆

Human nature guides social rules. As dictated by our evolution, most people most of the time in most circumstances are honest and fair and cooperative and want to do the right thing for their community and society. But most people are also competitive, aggressive, and self-interested and want to do the right thing for themselves and their family. This evolved disposition sets up two potential areas of conflict: within ourselves, our selfish desire for self-improvement conflicts with our altruistic desire for social enhancement, and our competitive desire to better our lot in life sometimes comes into conflict with the same desire that others have in themselves.

These potential sources of conflict mean that we need a society based on the rule of law and formal institutions. One of the first social institutions was the system of private property, and of course it has an evolutionary basis in the natural propensity for animals to mark their territories and defend them through threat gestures and even physical aggression. Such animal territories are the declaration of a private ownership of what was once a public good. Not all animals do this, and those that do are not cognizant that by marking their territories they are making a declaration of ownership. Yet we can see in this example evidence for the evolution of a premoral imperative for acquisitiveness and possessiveness, not unlike the moral emotions of empathy, revenge, and guilt seen in chimpanzees, capuchin monkeys, and other nonhuman primates.

Indeed, once a territory is declared taken by an animal, would-be trespassers have to invest considerable energy and risk grave bodily injury in attempts to acquire the property for themselves. There is an *endowment effect* such that we are more willing to invest in defending what is already ours than we are in trying to take what is someone else's. Dogs, for example, will invest more energy in defending a bone from a challenger than they will in absconding with some other dog's bone. The endowment effect with property ownership has a direct and obvious connection to loss aversion, where we are twice as motivated to avoid the pain of loss as we are to seek the pleasure of gain. Evolution has wired us to care more about what we already have than what we might possess, and this evolved moral emotion undergirds the concept of private property.

In the long history of humanity, as hunter-gatherers morphed into consumer-traders and populations expanded from hundreds to thousands to millions, personal ties of friendship were necessarily displaced by more formal webs of trust through social institutions that developed rules and reinforced them.

A problem that becomes apparent as soon as you think about this transition is that as the population of a community grows, the rules of how its members can live in relative harmony do not just grow in a linear fashion, in lockstep with the population. If two people in interaction represent a dyad, and three a triad, the expansion of a dyad into a triad requires the addition of not just one more rule, but many new rules. With each addition to the social equation, the number of rules needed to ensure social harmony and conflict resolution grows exponentially. This exponential growth in the potential for conflict is based on the fact that there are more people doing more things that may potentially conflict with what others want or need. From his studies of hunter-gatherer peoples in Papua New Guinea, for example, the UCLA evolutionary biologist Jared Diamond has shown what happens when you increase the population of a small hunter-gatherer group. In a band of, say, 20 people, there are 190 possible dyads, or two-person interactions (20 x 19 ÷ 2). But if a number of bands coalesce into a tribe of, say, 2,000 people, there will be 1,999,000 possible dyads (2,000 x 1,999 ÷ 2). The multiplier in population size between 20 and 2,000 is 100, whereas the multiplier in possible dyads between these two populations is a startling 10,521. And imagine what such a calculation would be for a corresponding increase in triadic interactions. Thus, the transition from bands and tribes of hundreds or thousands into chiefdoms and states of hundreds of thousands or millions represents a social leap so profound that it requires entirely new social technologies for communication, exchange, decision making, and conflict resolution. "Once the threshold of 'several hundred,' below which everyone can know everyone else, has been crossed, increasing numbers of dyads become pairs of unrelated strangers," Diamond explains. "Hence, a large society that continues to leave conflict resolution to all of its members is guaranteed to blow up. That factor alone would explain why societies of thousands can exist only if they develop centralized authority to monopolize force and resolve conflict."[12]

The centralized monopolization of force is, indeed, how most societies in history solved the problems that arrived with larger populations. In the Middle Land environment of our Paleolithic past, the evolution of reciprocal and indirect altruism led to the establishment of norms of reciprocity and the redistribution of food and other commodities among other members of the group. Such redistribution programs assure those who are successful in hunting or gathering on one day that they will not starve when they return home empty-handed on another day. Diamond applies his dyad calculation to the economies of redistribution, showing that "the same mathematics that makes direct pairwise conflict resolution inefficient in large societies makes direct pairwise economic transfers also inefficient. Large societies can function economically only if they have a redistributive economy in addition to a reciprocal economy. Goods in excess of an individual's needs must be transferred from the individual to the centralized authority, which then redistributes to the individuals with deficits."[13]

This conclusion makes sense from a folk economic perspective, where we simply scale up the social structure of hunter-gatherers into that of consumer-traders. In fact, societies have done just this throughout history. While it is true that most societies included top-down solutions in the form of chiefdoms, kingdoms, monarchies, theocracies, dictatorships, and the like, in the last half millennium the development of diverse economic and political institutions has flipped back to bottom-up solutions. And we have not simply increased the size and number of hunter-gatherer institutions and rules; we have added entirely new categories of similarly modeled social institutions, which in turn have created the need for entirely new categories of rules.

In particular, democratic capitalism offers a bottom-up solution based on the guaranteed right of private property and a democratic vote as well as the free and fair trade between citizens in the nation and with citizens in other nations. As a formal institution, it has developed a proven track record of resolving conflicts relatively peacefully (compared to its forerunners) in states and empires numbering in the hundreds of millions. The addition of the relatively new concept of property rights—which assigns the right and use of a commodity to someone with stipulated rules of conduct that are socially recognized and mutually binding among all members of the group regardless of the owner's ability to physically (and convincingly) threaten aggression—are counterintuitive to our evolutionary instincts and therefore require constant vigilance. Representative government allows everyone the right to participate in the construction of the rules of engagement and the social institutions that govern those rules and their enforcement, but this too has come about not naturally, but through the concerted efforts of social reformers going against the grain of deep time. Abraham Lincoln was right when he said in his first inaugural address in March 1861, on the eve of the most decisive and divisive war in our nation's history, "The mystic chords of memory, stretching from every battlefield and patriot grave to every living heart and hearthstone all over this broad land, will yet swell the chorus of the Union, when again touched, as surely they will be, by the better angels of our nature."[14]

11

DON'T BE EVIL

On Thursday, April 29, 2004, Americans awoke to one of the greatest shocks they had experienced since the fateful eleventh day of September 2001—the pictorial account of the abuses at the Abu Ghraib prison just outside Baghdad that revealed the dark underbelly of our side of the war in Iraq. There was the female pixie-cut American soldier in fatigues and a T-shirt dragging an Iraqi detainee down the hall on a leash. There was the pyramid of naked men, heads bowed in shame, with two American soldiers grinning in self-righteous triumph . . . thumbs up! There was the group of naked men milling about uncomfortably in a room, bags over their heads and hands feebly trying to cover their genitals. There was the faux fellatio scene between two Iraqi men. There was the terrified detainee about to be mauled by two unmuzzled Belgian shepherd dogs barely restrained by a couple of burly soldiers. And there was the now iconic crucifixion-by-electricity scene with the hooded man atop a box, arms outstretched and electrical wires wrapped about his neck, flowing from his hands, and disappearing upward to who knows where.

We all know that Other People are sadistic, cruel, and evil—after all, the "evildoers" are the ones we are fighting. But clean-cut, fresh-faced, football-playing, churchgoing, parent-loving, strapping young American evildoers? How was this possible? And who was minding the store? Isn't America an advanced democratic civilization based on the rule of law, the protection of civil liberties, and equal rights for all peoples? Our boys are not medieval inquisitors, Torquemadas in camouflage. How could this happen?

There was one American who wasn't shocked. In fact, he'd seen it all thirty-five years before in the basement of the psychology building at Stanford University where he conducted an experiment on the power of the environment to lead otherwise good people to become transitorily evil. When social psychologist Dr. Philip Zimbardo flipped on the news that night and saw the Abu Ghraib photographs, especially the head-bagged naked men in ersatz sexual acts, he experienced a flashback to the second week of August 1971, when he launched an experiment in which he randomly assigned a group of college student volunteers to be "guards" or "prisoners" in a mock prison environment. The experiment was to last two weeks. Zimbardo had to terminate it after six days when these intelligent, educated, wholesome, moral young men were transformed into cruel and sadistic guards or emotionally shattered prisoners. Had we learned nothing in three decades of research on the social psychology of evil, Zimbardo wondered? As he watched the parade of military leaders, politicians, and social commentators recoil in horror and proclaim that this was all the result of just a few bad apples, Zimbardo knew what he had to do. The result of this concatenation of past research and current events is what he calls *The Lucifer Effect* (also the title of his book), or the transformation of character that leads ordinarily good people to do extraordinarily evil things.[1]

Zimbardo argues that instead of attributing evil to a few bad apples, we should look more carefully at the bad barrels in which they are found. Evolution created in all of us the capacity for evil, but it is only rarely expressed, in very defined environments. When it comes to economic life, if Enron and the Gordon Gekko "Greed Is Good" ethic were the rule rather than the exception, market capitalism would have imploded long ago. Instead, Google and the "Don't Be Evil" corporate ethic is the rule.

For markets to be moral, there must be two conditions: (1) internal trust reinforced by personal relationships, and (2) external rules reinforced by social institutions. In the last two chapters we saw how trust is established and fortified through direct and personal interactions and strengthened and enforced through social institutions. In this chapter we shall see what happens when those institutional brakes fail.

The *dispositional theory of evil* holds that evil is the result of bad dispositions in people, whereas the *situational theory of evil* holds that evil is the product of corrupting circumstances. The dispositional theory of evil is the one most commonly embraced by religion (original sin), medicine (internal disease), psychiatry (mental illness), and the law (personal responsibility), whereas the situational theory of evil is more conventionally invoked by social psychologists, sociologists, and anthropologists sensitive to the power of environments to shape human behavior. But both of these theories can be simultaneously true and interactive. By disposition we have the capacity for good and evil, with the behavioral expression dependent on the situation and whether we choose to act. Consider the trenchant observation of Aleksandr Solzhenitsyn, who knew a few things about the capacity for evil that rests inside each of us, from his classic work *The Gulag Archipelago*:

> If only there were evil people somewhere insidiously committing evil deeds, and it were necessary only to separate them from the rest of us and destroy them. But the line dividing good and evil cuts through the heart of every human being. And who is willing to destroy a piece of his own heart?[2]

Our dual dispositional nature of good and evil arises from the fact that we evolved as a social primate species practicing within-group amity and between-group enmity. In order to survive as individuals we must get along with our fellow in-group members, and this led to the evolution of the moral sentiments of empathy, cooperation, and altruism—prosocial tendencies, or a good disposition. But the same evolutionary pressures that created a sense of attachment with our fellow in-group members fashioned a sense of animosity and xenophobia against out-group strangers, and this led to the evolution of the immoral sentiments of violence, competitiveness, and selfishness—antisocial tendencies, or an evil disposition.

An empirical demonstration of our natural inclination toward dividing the world into us versus them can be seen in a 1990 experiment by the social psychologist Charles Perdue. Subjects were told that they were participating in a test of their verbal skills in which they were to learn nonsense syllables, such as *xeh*, *yof*, *laj*, or *wuh*. One group of subjects had its nonsense syllables paired with an in-group word (*us*, *we*, or *ours*), a second group had its nonsense syllables paired with out-group words (*them*, *they*, or *theirs*), while a control group had its nonsense syllables paired with a neutral pronoun (*he*, *hers*, or *yours*). Subjects were then asked to rate the nonsense syllables on their pleasantness or unpleasantness. Remarkably (or maybe not, depending on your view of human nature), the subjects who had their nonsense syllables paired with in-group words rated them as significantly more pleasant than subjects who had their nonsense syllables paired with out-group words or neutral-paired words.[3]

We are good and evil. Most people in most situations are good and moral and do the right thing, but under certain circumstances the capacity for immoral and evil behavior that we all harbor in our hearts of darkness may be expressed.

✦ ✦ ✦

Philip Zimbardo was already a legend in psychology when I was in graduate school in the late 1970s, and his stature since has only grown. Born in the South Bronx and raised in poverty by his uneducated Sicilian parents, Zimbardo witnessed firsthand what can happen when people find themselves in environments with unenforced laws and openly expressed distrust. As a young professor at Stanford University, he set up an experiment in which he observed apparently abandoned vehicles—one in his old childhood haunt and one on the streets of upscale Palo Alto, California. In the Bronx, people started stripping down the car before the research team could finish setting up their hidden cameras. In just one day, there were twenty-three vandal attacks, all in the daytime and all but one by adults passing or driving by. In Palo Alto, by contrast, the car went untouched until Zimbardo finally gave up and drove the car back to campus, at which time three neighbors called the police to report that the car was being stolen. Such a striking difference cannot be the result of differential dispositions between New Yorkers and Californians. Quite obviously the difference between living in the Bronx and living in Palo Alto made all the difference. What were those differences?

This is the question Zimbardo set out to answer with the Stanford Prison Experiment, which now ranks alongside Sigmund Freud's couch, B. F. Skinner's boxes, and Stanley Milgram's obedience to authority experiments as seminal events in the history of science. If you've ever taken an introductory psychology course, you know Phil Zimbardo.

The details of the Stanford Prison Experiment are by now well known. In the basement of the psychology building a makeshift prison was constructed in which offices were converted into cells. The students randomly chosen to be prisoners were arrested at home by members of the Stanford Police Department, brought to the prison in squad cars, then sprayed for lice and forced to stand naked during orientation before they were finally given drab prison garb and crammed into six-by-nine-foot cells. For their part, the guards were given clubs, whistles, keys to the prison cells, and mirrored sunglasses ("an idea I got from the film *Cool Hand Luke*," Zimbardo noted in an aside). Over the next couple of days these psychologically well-adjusted American students were transformed into the role of either violent, authoritative guards or demoralized, impassive prisoners. The experiment was to last for two weeks. Zimbardo's girlfriend at the time (now his wife of over thirty years), after seeing the guards abusing the prisoners during their late-night toilet run, with bagged heads and chained ankles, insisted that Zimbardo end it before someone was seriously hurt. At that moment he realized that he had become part of the experiment in his role as prison superintendent. "I called off the experiment not because of the horror I saw out there in the prison yard," he explained in the technical write-up of the experiment, "but because of the horror of realizing that I could have easily traded places with the most brutal guard or become the weakest prisoner full of hatred at being so powerless that I could not eat, sleep, or go to the toilet without permission of authorities."[4]

I asked Zimbardo how he views the experience three decades later. "The message of my little Stanford Prison Experiment is that situations can have a more powerful influence over our behavior than most people appreciate and few people recognize," he began. "Social psychologists like myself have been trying to correct the belief that most people hold that evil is located only in the disposition of the individual—in their genes, their brains, their essence—and that there are good apples and there are bad apples." But there *are* bad apples, right? Yes, of course, Zimbardo concedes, but the vast majority of evil in the world is not committed by those few bad apples; instead, it is ordinary people doing extraordinary things under certain circumstances. Zimbardo prefers to err on the side of granting people the benefit of the doubt. "Before we blame individuals, the charitable thing to do is to first find out what situations they were in that might have provoked this evil behavior. Why not assume that these are good apples in a bad barrel, rather than bad apples in a good barrel?"

How can we tell the difference between good and bad apples, and between good and bad barrels? "When I launched my experiment at Stanford we knew these students were good apples because we gave them a battery of tests—personality tests, clinical interviews, we checked their background, etc., and every one of them was normal. Then we randomly assigned them to be guards or prisoners. So, on day one they were all good apples. Yet within days, the guards were transformed into sadistic thugs and the prisoners were emotionally broken." Zimbardo's bad barrel turned good apples rotten.

When the Abu Ghraib prison torture story broke, it wasn't long before Zimbardo, as well as the media, made the connection to the Stanford Prison Experiment. After appearing in several interviews, Zimbardo was contacted by one of the attorneys representing Staff Sergeant Ivan "Chip" Frederick, the military policeman in charge of the night shift on Abu Ghraib's Tiers 1A and 1B, the most abusive cellblocks in that most abusive of Iraqi prisons. Without denying Frederick's culpability in the abuses (Frederick admitted his own guilt), Zimbardo knew he wanted to explore the deeper environment that enabled the torture, abuse, and humiliation of the prisoners. According to Zimbardo, before Frederick went to Iraq, he was an all-American patriot, "a regular church-going kind of guy who raises the American flag in front of his home each day, gets goose bumps and tears up when he listens to our National Anthem, believes in American values of democracy and freedom, and joined the army to defend those values." After Frederick was charged in the abuses, Zimbardo arranged for a military clinical psychologist to conduct a full assessment of Frederick using a battery of tests. The psychological assessments indicated that Frederick was by all counts a perfectly normal guy, average in intelligence, average in personality, with "no sadistic or pathological tendencies." To Zimbardo, these results "strongly suggest that the 'bad apple' dispositional attribution of blame made against him by military and administration apologists has no basis in fact."

Generalizing his findings from both the Stanford Prison Experiment and Abu Ghraib, Zimbardo posits a triad of factors—the *Person*, the *Situation*, and the *System*—and how together and in interaction with one another they can lead a good person to do evil things. Here we see an integration of the dispositional theory of evil (the Person), the situational theory of evil (the Situation), and a third component Zimbardo only added after considering the incidents at Abu Ghraib—the larger context in which the person and situation coexist (the System). "When I was reading the reports about Abu Ghraib I wanted to know who creates these situations that enable evil," he recalled. "The system is the bigger barrel: legal, economic, historical, political forces that give these situations legitimacy. And most systems have a shield so that there is no transparency."

✦ ✦ ✦

No one joins a cult. People join a group whose cause they believe in but that gradually, almost imperceptibly, turns bad. In the same manner, no one hires on with a major corporation dreaming of the day that they can bankrupt the company and divest stockholders of their investments; meltdowns such as those at Enron and WorldCom begin with people doing something that they believe is for a good cause but that progressively, step by step, morphs into evil. If you sat people down in front of a bank of electrical switches and asked them to throw the final switch in the sequence that would deliver to an innocent person a lethal charge of 450 volts, almost no one would do it. But if you concoct a situation in which subjects are playing the role of a teacher in a memory experiment to test whether punishment helps or impedes learning, it is easy to get them to deliver a harmless little 15-volt shock, then a slightly noticeable 30-volt shock, and so on up the path to 450 volts of evil.

I am speaking, of course, about the Yale University social psychologist Stanley Milgram's famous 1960s experiments on obedience to authority, whose results directly support the point I am making here: that our dual propensities for good or evil can be dramatically tweaked one way or the other depending on the situation and the system. To more closely model the real world, instead of college students, Milgram's sub-jects were a thousand local townspeople from New Haven and Bridge-port, Connecticut. As he advertised in the local paper, he wanted "factory workers, city employees, laborers, barbers, businessmen, clerks, construction workers, sales people, telephone workers" who would par-ticipate in "a study of memory." The protocol called for the subject to read a list of paired words to the "learner" (a shill working for Milgram), then present the first word of each pair again, upon which the learner had to recall the second word. If the learner was wrong, the teacher was to deliver an electric shock. (No one was actually shocked, but the shill faked it and the subjects believed that the shocks were real.) The shocks were delivered by the subjects, who were seated before a box with toggle switches labeled *Slight Shock, Moderate Shock, Strong Shock, Very Strong Shock, Intense Shock, Extreme Intensity Shock, DANGER: Severe Shock, XXXX* ("the pornography of power," Zimbardo calls this final level of 450 volts).

Before the experiment, Milgram surveyed forty psychiatrists and asked them to predict what percentage of subjects would go all the way to the 450-volt level. The answer consistently given by these professional observers of human behavior was 1 percent—only sadists would engage in such sadistic behavior. Wrong! These judges of human nature were op-erating under the traditional dispositional model of locating all pathologi-cal actions within the head of pathological characters, ignoring the situational demands put upon the subjects. The answer consistently given by the experiments was that 65 percent of subjects went all the way to XXXX, zapping their would-be learners with a lethal dose of 450 volts! If that isn't appalling enough, every last one of Milgram's subjects went as far as the Strong Shock level of 135 volts. Such is the power of the situa-tion to push even dispositionally good people to commit evil deeds.[5]

So disturbing were the initial results that Milgram ran more experi-ments, varying the conditions to control for other intervening variables that might have skewed his initial findings. Additional research con-firmed and refined the initial conclusions about evil. The physical prox-imity of the victim, for example, influenced how far the teachers would go in shocking the learners—the closer they were in the experiment, the less shock was delivered; the farther away they were, the more shock was administered. Not surprisingly, group pressure was also a factor, as when Milgram had additional confederates encourage the teacher to go further up the line of electric shocks; but when the confederates disobeyed Milgram's authority and refused to continue the experiment, the actual subjects were also inclined to disobey. In other words, Milgram discovered that good and evil are far more dependent on the situation than they are on disposition. These findings support my evolutionary model of evil, in which social pressures from one's fellow in-group members are especially forceful in shaping behavior.

As disturbing as these findings are, in many respects Zimbardo's experimental results are even worse. In Milgram's experiments an authoritative figure is standing over the subjects commanding them to continue the experiment and reminding them of their obligation as participants in this important research, whereas Zimbardo found that subjects turned evil just by the circumstances of the setup and needed no additional encouragement once they were assigned to their respective roles as guards and prisoners. This is what Hannah Arendt really meant by *the banality of evil*.

Because we evolved to be such social beings, we are hypersensitive to what others think about us, and we are strongly motivated to conform to the social norms of our group. Solomon Asch's studies on conformity demonstrate that groupthink is so powerful that if you are in a group of eight people instructed to judge the length of a line by matching it to three other lines of differing lengths, even when it is obvious which line is the match, if the other seven people in the group select a different line, you will agree with them 70 percent of the time. The size of the group determines the degree of conformity. If it is just two people judging the line lengths, conformity to the incorrect judgment is almost nonexistent. In a group of four in which three select the incorrect line match, conformity happened 32 percent of the time. But no matter what the size of the group, if you have at least one other person who agrees with you, conformity plummets.[6]

Not surprisingly, an fMRI study even identified which parts of the brain light up under conformity versus nonconformity. Conducted by Emory University neuroscientist Gregory Berns, the task involved matching rotated images of three-dimensional objects to a standard comparison object. Subjects were first put into groups of four people, but unbeknownst to them, the other three were confederates who would intentionally select an obviously wrong answer. On average, subjects conformed to the group's wrong answer 41 percent of the time, and when they did, the areas of their cortex related to vision and spatial awareness became active. But when they broke ranks with the group, their right *amygdala* and right *caudate nucleus* lit up, areas associated with negative emotions.[7] In other words, nonconformity is an emotionally traumatic experience, which is why most of us don't like to break ranks with our social group norms.

We all like to think that we are immune to peer pressures and social stresses, but the distressing conclusion from Milgram's studies is that as a social primate species we are remarkably susceptible to the wishes of others, especially alpha males and those in positions of authority. "What is surprising is how far ordinary individuals will go in complying with the experimenter's instructions," Milgram recalled. "It is psychologically easy to ignore responsibility when one is only an intermediate link in a chain of evil action but is far from the final consequences of the action." The combination of a stepwise path and an authority pushing you to continue at each step is what makes most evil of this nature so insidious. Milgram broke the process down into two stages: "First, there is a set of 'binding factors' that lock the subject into the situation. They include such factors as politeness on his part, his desire to uphold his initial promise of aid to the experimenter, and the awkwardness of withdrawal. Second, a number of adjustments in the subject's thinking occur that undermine his resolve to break with the authority. The adjustments help the subject maintain his relationship with the experimenter, while at the same time reducing the strain brought about by the experimental conflict."[8]

When order breaks down, when the rules are no longer enforced, when the normal institutional brakes on evil are lifted, evil is facilitated through the contagious excitement of the group's actions, through the unchecked momentum of the smaller bad steps that came before, and ultimately permission for evil is granted by the system at large. This is what happened on November 18, 1978, in the jungles of Guyana when Jim Jones, leader of the People's Temple cult, which had moved its operations there from the more transparent and rule-bound culture of San Francisco and Los Angeles, ordered the mass suicide and genocide of his own followers by inducing them to drink a cyanide-laced Kool-Aid drink. Although some members did try to escape (and were shot), and some members had to be forced to drink the poison, most got caught up in the contagion of the moment.

◆ ◆ ◆

In the end, of course, people choose to be good or evil. We can change the conditions and attenuate the potential for evil, first by understanding it and then by taking action to change it. As the nineteenth-century British statesman Edmund Burke warned, "The only thing necessary for evil to triumph is for good men to do nothing."[9] Doing nothing, unfortunately, is the default option in such circumstances. Evolution programmed us to protect our kin and kind and not risk our lives by standing up to evil. To do so requires not only a heroic set of circumstances, but a heroic disposition. It takes a hero to stand up to evil. The whistleblower at Abu Ghraib, Joe Darby, is a hero. Enron vice president Sherron Watkins is also a hero for her early warnings about the imminent collapse of the company. And WorldCom internal auditor Cynthia Cooper was a hero when she blew the whistle on the fraudulent accounting practices there that kept $3.8 billion of losses off the books.

In what way do corporate environments enable evil? I put the question to Zimbardo. "Corporations are a series of situations, especially when they get really big. There's a situation in the corporate board office, a situation in the accounting office, a situation in the public relations office, and so forth. In each of those settings there are sets of norms that tell employees what they have to do to be a team player. This sets up an in-group/out-group situation, and if you want to get bonuses and raises and promotions you have to be an in-group team player and go along with the norms in that situation." Right, but who gives out those raises and promotions, I asked? And isn't there a corporate philosophy of good or evil that comes down from the top? "Yes, and if the head of the team or group or company has a certain ideology, you have to follow it in order to get ahead. Perhaps that ideology is a little different from what you thought it would be, but it's usually a small step to take, and once you've invested into being on the team it is hard not to go along with it. It would be dissonant to give up all you've done not to take one more small step down the road."[10]

Earlier in the book we discussed market path dependency, in which markets become dependent on the paths they are already on, or become locked into the channels in which they are operating. Here we find an example of moral path dependency, in which moral systems and behavior become dependent on the rules of the corporate environment, or become locked into the channels of moral patterns exhibited by others in the environment. Thus, an environment of moral corporate philosophy and leaders establishes a situation that can either accentuate the good disposition of employees or bring out the bad. Let's examine these two corporate systems, one evil and the other good, and see how their respective environments led to vastly different outcomes.

In the 1987 film *Wall Street*, Michael Douglas's character, the high-rolling corporate raider Gordon Gekko, explains to Charlie Sheen's youthfully naïve Bud Fox, "We make the rules, pal. The news, war, peace, famine, upheaval, the price of a paper clip. We pick that rabbit out of a hat while everybody sits around wondering how the hell we did it. Now you're not naïve enough to think that we're living in a democracy, are you, Buddy? It's the free market, and you're part of it." Echoing the myth about evolutionary theory that most people hold—that nature is "red in tooth and claw" (in Tennyson's memorable phrase) and evolution is nothing more than the "survival of the fittest" (in Herbert Spencer's unfortunate characterization)—Gekko explains why America has lost its standing atop the industrial world: "The new law of evolution in corporate America seems to be survival of the unfittest. Well, in my book you either do it right or you get eliminated."

This is the generally received view that most people hold about most of corporate America and market capitalism; in this perspective, the Gordon Gekkos of the world do not symbolize the occasional CEO bad apple but the entire rotting corporate barrel. In one short excerpt from Gekko's now-famous "greed" speech we find all the myths that have been exploded in this book: that capitalism is grounded in and depends upon cutthroat competition, that businessmen must be self-centered

and egotistical to achieve success, that evolution is selfish and only win-
nows and never creates, and, of course, that greed is good:

> The point is, ladies and gentlemen, that greed—for lack of a better
> word—is good. Greed is right. Greed works. Greed clarifies, cuts
> through, and captures the essence of the evolutionary spirit. Greed,
> in all of its forms—greed for life, for money, for love, knowledge—
> has marked the upward surge of mankind. And greed—you mark
> my words—will not only save Teldar Paper, but that other malfunc-
> tioning corporation called the USA.

Do people really believe that this description best characterizes the
corporate world? They do. A representative slice of the beliefs many people hold about corporations can be seen in the 2003 Canadian documentary film *The Corporation*, whose subtitle tips its hand: *The Pathological
Pursuit of Profit and Power*.[11] Noam Chomsky, Howard Zinn, Jeremy
Rifkin, Michael Moore, and other anti–big business public intellectuals
liken corporations to vultures, Godzilla, Frankenstein's monster, and
other beastly creations "trying to devour as much profit as possible at anyone's expense." Chomsky equates corporations to slavery, implying that
the very concept of a corporation should be unconstitutional. The psychologist Robert Hare applied the Diagnostic and Statistical Manual (the
DSM-IV) of the American Psychiatric Association to diagnose corporations as suffering from clinical psychopathy, implying that corporations
are diseased and should be institutionalized for mental illness. Even
many corporate insiders buy into the myth that market capitalism is
founded on and sustained by selfish, cutthroat, every-man-for-himself ra-
pacity. For example, the corporate raider Ivan Boesky, after whom the
Gordon Gekko character is said to have been patterned, during a 1985
commencement speech at the University of California, Berkeley, told students: "Greed is all right, by the way. I want you to know that. I think
greed is healthy. You can be greedy and still feel good about yourself."[12]

Yet if biological evolution in nature was really founded on and sustained by nothing more than a winner-take-all understanding of greed,
life on earth would have been snuffed out hundreds of millions of years
ago; if market capitalism was winner-take-all, it would have collapsed
centuries ago. This is, in fact, why WorldCom and Enron type disasters

still make headlines. If they didn't—if such corporate catastrophes caused by egregious ethical lapses were so common that they were not even worth covering on the nightly news—free market capitalism would go the way of the dinosaurs.

This is not in any way to deny the abuses committed by corporations during the past two centuries, or to paint a rosy portrait of corporations that has no basis in reality. My point, however, is that such abuses are the exception, not the rule; they are the product of evil corporate cultures more than they are the result of evil corporate leaders. The contrast between Enron and Google, then, will serve to make the contrast in striking detail that demonstrates the power of the situation to draw out the better or worse angels of our natures.

✦ ✦ ✦

The story of Enron's collapse has been well recounted in excruciating detail in a number of popular books, scholarly articles, and a documentary film, most of which attribute the company's demise to accounting irregularities and corrupt individuals at top levels of management.[13] Echoing his description of what happened at Abu Ghraib, in fact, President George W. Bush attributed the company's downfall to a few "bad apples." The "bad apples" theory, however, does not explain what happened at Enron, nor does it give us any deeper insight into the true nature of corporate evil. In a comprehensive study of the evolution of Enron's corporate culture, management analysts Clinton Free and Norman Macintosh, from the Queen's University business school, found that something happened between the time of Richard Kinder's term as president from 1986 to 1996, when Enron operated with a highly effective managerial system that included transparent governance practices, to the Jeff Skilling era from 1996 to 2001, in which such checks and balances were neutralized.[14] What was it that happened?

Enron began in 1985 when Kenneth Lay orchestrated the merger of the Houston Gas Company with Internorth Inc., becoming CEO of the new energy corporation. Lay then hired Richard Kinder to run it for him while he brokered deals and curried political favors in Washington. "Kinder ran the company . . . while Lay gave speeches and posed for pictures" is how one insider described it. From 1990 to 1996, Enron's

reported earnings increased from $202 million to $584 million, while its revenues skyrocketed from $5.3 billion to $13.4 billion.[15]

The key to Kinder's management style was transparency, accountability, and involvement at every level of the company. At regular meetings with managers and department heads, Kinder expected everyone to come prepared to be grilled in great detail about every aspect of their job, and with a near photographic memory, Kinder was not easily fooled. As one manager later remembered, "You could give him a budget number and explain where it came from and he'd say, 'That's not what you told me last year.' And then he'd go to his desk and retrieve the year-earlier budget and prove you wrong. It was amazing." Another unit leader said that Kinder "was impossible to bullshit" and if managers "lied to him about their numbers, Rich would eat them for lunch."[16]

Evil often happens in hidden places—the basement of the Stanford psychology building, the deep recesses of the Abu Ghraib prison. The first line of defense against evil, then, is transparency, open communication, and the constant surveillance of every aspect of a system. In regular face-to-face meetings with his managers, Kinder—known as "Doctor Discipline" at Enron—demanded up-to-the-minute reports such that he always knew who was doing what to whom and when. As one Enron executive of twenty years recollected, "Kinder would sit in that room with his yellow pad and he knew every god-damned thing happening in that company."[17] Free and Macintosh note that this style of management closely follows the model outlined by the Harvard Business School professor of management, Robert Simons, in his influential book *Levers of Control*, in which he identifies three control systems that allow strategic and ethical change: (1) belief systems that communicate core values and provide inspiration and direction, (2) boundary systems that frame the strategic domain and define the limits of freedom, and (3) interactive systems that provide flexibility in adapting to competitive environments and encourage organizational learning.[18] Kinder employed this management system by closely reading his managers' reports, then challenging and debating them at regularly scheduled face-to-face meetings; in turn, he had these managers do the same thing with the people under them, so that at every level Enron was transparent and thus less susceptible to mismanagement and corruption. Further, Kinder fostered a family-like

atmosphere at Enron, for example, showing care and concern for the personal lives of his employees (such as paying the travel expenses for one of his managers to return home for a family funeral), which tends to engender respect and loyalty.

All this changed in 1997, however, when Jeff Skilling replaced Kinder as president. A graduate of Harvard Business School and a fan of Richard Dawkins's epochal book *The Selfish Gene*, Skilling misread the theory to mean that evolution is driven by cutthroat competition and self-centered egotism, and he liked the notion of "the survival of the fittest," which led directly to a policy he implemented at Enron called the Peer Review Committee (PRC) system, known in the company as "Rank and Yank." PRC was based on the mistaken presumption that people are motivated primarily by greed and fear, which Skilling drove by ranking employees on a scale of 1 to 5, with 5's being given the boot. This resulted in 10 to 20 percent of his employees being axed every six months, leaving everyone on edge and in a state of anxiety over job security. The formal reviews were posted on a company Web page along with a photograph of the employee, increasing the potential for personal humiliation. Those who received a 5 in the relative ranking system—no matter how good their absolute performance may have been—were automatically sent to "Siberia," a purgatory where the 5's had two weeks to find another position at Enron, after which they were "out the door."

As Ken Lay described it, "Our culture is a tough culture. It is a very aggressive culture." Charles Wickman, one of Enron's energy traders, described the corporate culture under Skilling this way: "If I'm on my way to the boss's office to argue about my compensation, and if I step on somebody's throat on the way and that doubles it, well, I'll stomp on the guy's throat. That's how people were."[19] While he was producing his 2005 documentary film on Enron, director Alex Gibney uncovered a cache of audiotapes from a West Coast energy company in which Enron traders can be heard requesting power station engineers to manufacture the shutdown of energy stations in order to decrease energy supplies along a particular grid, thereby boosting energy prices, from which Enron directly benefited. In 2000, this led to rolling blackouts in California, significant increases in energy bills, and, of course, a huge spike in Enron's stock price. When fire season exploded in California, further disrupting the energy grid and driving prices through the ceiling, one trader can be heard excitedly saying "Burn, baby, burn!"[20]

Skilling's evaluation and bonus system led to a lot of behind-the-scenes wheeling and back-door dealing between department heads and managers, swapping review evaluation points like so much horse-trading. Here is one typical conversation recounted by a manager: "'I was wondering if you had a few minutes to talk some PRC.' She replied, 'Why—you want to cut a deal?' 'Done,' I said—and just like that we cut our deal."[21] Another manager described the PRC system as creating "an environment where employees were afraid to express their opinions or to question unethical and potentially illegal business practices. Because the rank-and-yank system was both arbitrary and subjective, it was easily used by managers to reward blind loyalty and quash brewing dissent."[22] By pitting employees against one another, the rank-and-yank system established an environment that brought out the worst in Enron's employees: selfishness, competitiveness, and greed.

In addition to his belief in the outdated and untenable doctrine of social Darwinism, Skilling was a high risk taker—short on dopamine, we might conjecture—driving him to take ever greater risks with both his body and his company. Adventurous corporate trips, such as a motorcycle expedition down the rugged terrain of Baja California, reinforced the macho competitive atmosphere of Enron's corporate environment. Skilling's bonus system, based on the PRC database rankings in which employees were arrayed on a bell curve, further eroded any sense of team spirit. Because bonuses ranged from 10 to 26 percent of an employee's take-home pay, there was considerable motivation to manipulate the ratings in order to boost the rankings in the hierarchy, as well as to backstab each other and sabotage deals put together by other employees and departments. One executive said that the bonus system "had a hard Darwinian twist" that made "a humongous difference on Enron by instilling a competitive streak in every employee." Another noted: "It's a Jeff Skilling thing. It's his baby that he created to keep us on our toes . . . or on edge, anyway."[23] Another division head noted how the system led to secrecy and suspicion within the company, which fostered deception about owning up to the company's actual state of financial health when it subsequently got into trouble: "Every division and business unit was like its own silo, separate from all the other businesses. It was decentralized and not heavy on teamwork, with all of the divisions in competition with each other for resources. But since most only saw their part of the business, they assumed the problems were isolated. You understood your piece of the business and maybe what the guy next to you did, but very few understood the big picture."[24] Ultimately, what causes corporate corruption is an environment of evil established by the founders, executives, directors, and managers within a corporation—in short, its corporate social psychology—which then creates situations that encourage our hearts of darkness to beat faster.

✦ ✦ ✦

In contrast to the Gordon Gekko theory of economics that generates a bad-barrel corporate environment that can readily turn good apples into bad, the Google Guys theory of economics produces a good-barrel corporate environment that optimizes the good-appleness of its employees and customers.

I first met Sergey Brin and Larry Page at a weekend gathering in Seattle for gifted high school students called Adventures of the Mind, featuring a veritable Who's Who from the worlds of art, literature, science, and business brought in to hang out with and enthuse aspiring youngsters, who for three days could find themselves in casual conversation with a Nobel laureate, a Pulitzer Prize winner, a MacArthur "genius," or the creators of the world's most popular search engine. Brin and Page were almost too good to be true. With the wide-eyed wonder of new college grads about to embark on a mission to change the world, they spoke of colonizing space and cataloguing all the world's information with the same casualness that other people talk about a driving vacation or the latest book they read. After walking across the stage on his hands, Brin recounted the now-familiar story of how he and Page hatched the idea for their search engine in their Stanford University dorm room, how they got started by stripping unclaimed campus computers of their hard drives and stacking them one atop the other on racks, how the heat from so many computers running simultaneously created a serious fire hazard, how Stanford University declined an offer to purchase the search system, how the two Ph.D. students dropped out of school to pursue their search engine project, and how an investor gave them a check for $100,000 to open an office, but they couldn't deposit it until they created a company account, which they couldn't do until they concocted a company name. A googol is a number equal to 1 followed by a hundred zeros (or 10^{100}). It's a huge number, the sort of number that brings to mind the amount of information in the world to be indexed by the boys' search engine, and since Page was fascinated with mathematics, the guys chose that for their company name. But, as Brin joked, since they had yet to invent their spell-check algorithm, Googol was transformed into Google. One of the fastest-growing and most influential companies in corporate history began with a spelling error.

With their characteristic playfulness, Brin and Page named their Mountain View, California, corporate headquarters the Googleplex. If a googol is a 1 followed by 100 zeros, or 10^{100}, a googolplex is a 1 followed by a googol of zeros, or $10^{(10^{100})}$, or, in the words of the concept's inventor, nine-year-old Milton Sirotta, "a one, followed by writing zeroes until you got tired." That number too seemed appropriate when I visited the Googleplex, since everything about the place seems positively Brobdingnagian. When you enter the lobby you encounter lava lamps, a piano, and a giant global projection of a rotating earth with the numbers of search queries represented by rays of colored light beamed into space. A giant white board called "Google OS" (Operating System) is chockablock with multicolored Expo Marker–produced geek-o-centric flowchart goals for the company, such as *Develop AI*, *Orbital Mind Control*, *Google Football League (GFL)*, *Buy New Zealand*, *Build Singularity*, *Crop Circles*, *Childcare Kinderplex*, and, appropriately, *Elimination of Evil*. It is toward this latter goal that the Google environment is structured, starting with its corporate slogan, "Don't Be Evil."

My point in describing this environment is to contrast it with the environment of evil that leads people down the wrong path. Environments are both physical and psychological, and the Google lobby sets the tone for what awaits inside the glass doors. Speaking of which, glass doors and walls are transparent, and transparency is one of the foundations of trust. Transparency is apparent even in what millions of people from around the world are searching for, in an electronic board that displays scrolling search queries (presumably screened for X-rated requests).

Hallways contain bicycles and large rubber exercise balls. Googlers—as employees are known—work in small group clusters, sharing space with couches and dogs.[25] Googlers work hard because they play hard, and the Google campus is loaded with workout rooms, video games, foosball tables, pool tables, Ping-Pong tables, volleyball courts, and assorted other recreational conveniences. And if all that were not enough to make employees think twenty-seven times before pilfering pens and Post-it notes or embezzling checks and click-ad funds, free meals are available at assorted on-site restaurants, and numerous snack bars offer a variety of goodies to munch on between meals. Professional chefs prepare healthful and delicious food, which nine out of ten employees cited as what they most liked about their jobs. "There is such a thing as a free lunch after all," reads a Google Web page description of the top ten reasons to work there. "In fact we have them every day: healthy, yummy, and made with love."

Of course, all economists know that there is no free lunch. The business model to justify feeding thousands of people a day is as obvious as it is logical: feeding your employees means that they will not leave the Googleplex grounds for lunch and dinner, and will thus spend more time on campus (as tech companies are wont to call their headquarters) and less time driving, parking, and eating somewhere off the property. Or taking care of laundry, going for a haircut, getting their car washed, or enjoying a massage, all of which can be done at the Googleplex. Google even charters buses equipped with wireless Internet access so that employees who commute from San Francisco can be more productive. It is an environment that fosters a sense of both teamwork and independence. "People talk over lunch about the things they are playing with," noted one Google software engineer. "It is like they are the CEO of their own little company."

There is another reason for offering employees free food and convenient amenities: *reciprocity*. The fundamental principle of reciprocity evolved in its most base form as food sharing among primates, and has since developed into complex networks of exchange employed by everyone from mass-mailing merchants to Madison Avenue marketers: if I give you something for free, you will feel obligated to reciprocate.[26] Hunter-gatherer groups accumulate psychic credit with other groups by throwing a feast (for example, the Native American potlatch), which must be paid back in kind to maintain political capital, build economic trust, and generate social goodwill. Consumer-traders accumulate psychic credit by throwing the equivalent of a potlatch, which must be reciprocated in kind to maintain political, economic, and social equilibrium. Give a small gratis gift to potential customers and you increase your chances of turning them into actual customers. Readers my age and older will recall Jehovah's Witnesses handing out flowers at airports (no longer allowed) in hopes of guilting people into making a donation. More recently, a common strategy by mass marketers is to include a small gift, such as envelope stickers with your name and address on them, on the assumption that you'll use them and then feel the need to reciprocate by placing an order. One of the more blunt instruments of reciprocity I've seen is by pollsters who include a crisp new dollar bill with the survey they hope you will then complete.

The Google environment accentuates amity and attenuates enmity by minimizing corporate hierarchy and maximizing cross-pollination between people in different departments. "Because everyone realizes they are an equally important part of Google's success, no one hesitates to skate over a corporate officer during roller hockey," explains a statement on corporate culture that employees are encouraged to read.[27] Googlers are even expected to devote 20 percent of their time to exploring new ideas and projects, without hierarchical supervision. A horizontal corporate structure generates an atmosphere of egalitarianism and nonelitism that taps into our folk economic intuitions, and that atmosphere expands beyond the Googleplex and throughout the world. "Google's goal is to provide a much higher level of service to all those who seek information, whether they're at a desk in Boston, driving through Bonn, or strolling in Bangkok," reads another Web document on Google philosophy.

Consider the implications of the Google Books Library Project, in which millions of books from the libraries at Stanford, Harvard, Oxford, the University of Michigan, the New York Public Library, and others are being scanned and made available online, for free, and searchable by anyone from anywhere in the world. There are copyright issues with this project still to be resolved, of course. But my point is that projects like this reinforce an environment of trust and are thus an important step in the millennia-long march toward greater freedom and prosperity for more people in more places. As Brin and Page wrote in their document released with the company's initial public offering, "We believe a well functioning society should have abundant, free and unbiased access to high quality information. Google therefore has a responsibility to the world." Those who control information control the world, but if everyone has access to that information, no one can control the world. Information transparency trumps political hegemony.

From the user's perspective, none of this corporate environment matters. We just want a search engine that delivers what we need. Google's success has come in part from the fact that they actually deliver on the "customer first" maxim. If you build it they will come, and come they have, in droves, almost entirely by word of mouth (Google does next to no advertising of its brand). Why? Because users don't feel that corporate profits come first. Although Google could sell placement in search results, they don't, building trust capital with users who have come to know that what appears on the screen is the product of a search engine algorithm based on the most democratic of principles: the Google PageRank program determines which sites have been voted the best sources of information by the millions of other sites interested in that same information in the democratic world of cyberspace. "Our users trust Google's objectivity and no short-term gain could ever justify breaching that trust." And unlike most Web pages that try to monopolize their users' time, Google's stated goal "is to have users leave its website as quickly as possible."

Even what few ads there are on search pages (but never the home page, which remains in its minimalist mode) are just another source of information. Expanding the company motto, Google's clickthrough advertising program means "you can make money without doing evil." And Google advertising is as democratic as its search algorithm, because anyone can participate, no matter how small. Who needs an advertising agency to spend millions on focus groups and brand tests when you can run the experiment yourself for pennies on the dollar?

The central pillar of Google's code of conduct is its now familiar slogan, "Don't Be Evil."[28] But what does this clichéd phrase really mean? "It means making sure that our core values inform our conduct in all aspects of our lives as Google employees."[29] And what are those core values? Brin and Page's answer is the very model of a statement of how markets can be moral when they are grounded in a foundation of trust: "Being a Googler means holding yourself to the highest possible standard of ethical business conduct. This is a matter as much practical as ethical; we hire great people who work hard to build great products, but our most important asset by far is our reputation as a company that warrants our users' faith and trust. That trust is the foundation upon which our success and prosperity rest, and it must be re-earned every day, in every way, by every one of us."

The code of conduct goes on for pages detailing all manner of potential evils to avoid, for example, dealing with competitors' private information. Here we see the reign of the golden rule: "The level of business ethics to which we aspire requires that we apply the same rules to our competitors' information as we do to our own, and that we treat our competitors as we hope they will treat us. We respect our competitors and, above all else, believe in fair play in all circumstances; we would no sooner use a competitor's confidential information to our advantage than we would wish them to use ours. So, although gathering publicly available information about competitors is certainly a legitimate part of business competition, you should not seek out our competitors' confidential information or seek to use it if it comes into your possession. If an opportunity arises to take advantage of competitors' confidential information, remember: don't be evil. We compete, but we don't cheat."[30]

Of course I am well aware of the controversies that have arisen with Google's growth, including click fraud, the use of competitors' trademarked keywords in Google's AdWords advertisements, the inclusion of morally questionable content in Google Groups (most notably pornographic content and racial hate speech), copyright issues associated with Google's purchase of YouTube, and the high-profile case of Google in China, in which the company was forced to make concessions for the censorship of politically sensitive material in order to gain access into the country. Controversies of this nature, of course, are inevitable for any company that grows as rapidly as Google has, and no matter how lofty a company philosophy may be, perfection will always be an unattainable goal.

"Don't Be Evil" is a moral standard toward which to aim, not a sinless existence whose unachievability means no such norm should be invoked. The point of having moral codes—whether you are a hunter-gatherer or a consumer-trader—is to construct an environment of trust that encourages the expression of moral behavior.

12

FREE TO CHOOSE

In one of his more trenchant observations on the human condition, Thomas Jefferson wrote, "Freedom is the right to choose, the right to create for oneself the alternative of choice. Without the possibility of choice, and the exercise of creation, a man is not a man, but a member, an instrument, a thing." Given all the biological, psychological, and evolutionary baggage we carry into the marketplace, are we really free to choose?

Science has made a first line of inquiry into decision making and choice behavior in a new field called *computational neuroscience*, pioneered by the neuroeconomist Read Montague and discussed in his improbably titled work, *Why Choose This Book?*[1] Computational neuroscience focuses on the information processing that goes on inside our brains. Montague argues that our brains evolved computational programs to evaluate choices in terms of their value and efficiency as judged by goals that we care about. *Value* gives a choice option a computational number. *Efficiency* determines whether the choice is worth making. Those organisms that correctly compute the costs and long-term benefits of their choice options will be more efficient than those that do not, and as a consequence they are more likely to survive and pass on their genes for making good decisions. We are their descendants.[2]

Life, like the economy, is about the efficient allocation of limited resources that have alternative uses, to paraphrase the economist Thomas Sowell. It all boils down to determining the value of the available choices and what each one costs in terms of energy efficiency. To a predator, says Montague, prey animals are packages of energy to be consumed in order to recharge the computer's batteries. "This doctrine mandates that evolution discover efficient computational systems that know how to capture, process, store, and reuse energy efficiently," Montague suggests. Those that do, pass on their genetic programs for efficient computational neural processing to make efficient choices. Over the course of millions of years our brain has evolved to be so efficient that it consumes about one-fifth of the energy of an average lightbulb, costing about a nickel a day to run.[3] That's cheap! And cool. Think of how hot your computer gets even while it is just idling. If our brains were designed like our computers, our heads would be too hot to touch.

According to the *Computational Theory of Mind* (CToM), the mind is nothing more than a computational program running on a special type of machine called the brain. "It's the information processing that the brain carries out that is equivalent to our thoughts, not the parts themselves," Montague explains.[4] The reason that our computational programs can determine the value of a choice is that we evolved the capacity to care, and more precisely, to care about one option more than another. Our computers can run computational programs, but they cannot do what the brain's computational programs can do, which is to *care about* the choices made.[5]

Computational programs are designed by evolution to learn how to solve certain tasks, and society then tweaks these ancient programs toward specific cultural preferences. Rats, for example, inherit programs that are especially good at learning mazes and pressing bars because they evolved to forage in dark and spatially complex environments. Humans inherit programs that are especially good at visual acuity tasks and at navigating social situations because we evolved to forage in trees and to negotiate complex social environments. There are no blank slates for mice or men. Species have different goals, but the deeper underlying purpose behind goal-seeking behavior is the same. "Despite their differences," Montague continues, "all goals have one thing in common: They can all be used by our brains to direct decisions that lead to the satisfaction of the goal."[6]

Unfortunately, these evolved computational programs can be hijacked. Addictive drugs, for example, rewire the brain's dopamine system—normally used to reward choices that are good for the organism, such as food, family, and friends—to reward choosing the next high instead. We have all marveled at the inanity of rich and successful athletes, actors, artists, and others who throw away their careers, lose their friends, and finally abandon their families, all in exchange for recharging their dopamine receptors one more time . . . to the point where they end up destitute, incarcerated, or dead. How can this be? Are brain chemicals really that powerful? They are.

Here's how the system works . . . and doesn't work. In the brain stem—one of the most evolutionarily ancient parts of the brain, shared by all vertebrates—there are pockets of roughly fifteen thousand to twenty-five thousand dopamine neurons on each side of the divided brain that shoot out long axons connecting to other parts of the brain. These neurons stimulate the release of dopamine whenever it is determined that a received reward is more than expected, thereby causing the individual to repeat the behavior. Thus, the release of dopamine is a form of information, a message that says "Do that again." (Recall that the dopamine system is probably more of a *wanting system* than it is a *liking system*.) Addictive drugs take over the role of reward signals that feed into the dopamine neurons. So too do addictive ideas, most notably addictive *bad* ideas, such as those propagated by cults that lead to mass suicides (in the case of Jonestown and Heaven's Gate), or those propagated by religions that lead suicide bombers to commit mass murder (in the case of Islamic militant extremists).

I have made the case in this book that we evolved moral emotions that operate similarly to other emotions, such as hunger and sexual appetite. Thinking of these emotions as proxies for highly efficient computational programs deepens our understanding of the process. When we need energy, we do not compute the relative caloric values of our food choices; we just feel hungry for certain food types, eat them, and are rewarded with a sense of satisfaction. Likewise, in choosing a sexual partner, the brain employs a computational program to make you feel attracted to people with good genes, as indicated by such proxies as symmetrical face and body, clear complexion, and the hourglass figure in women and the inverted-pyramid build in men. In a similar manner, when we make moral choices about whether to be selfish or altruistic, we experience the emotion of guilt or pride for having done the wrong or right thing, but the moral calculation of what is best for the individual and the social group was made by our evolutionary ancestors. Emotions such as hunger, lust, and pride are stand-ins for these computations. As Blaise Pascal famously concluded, "The heart has its reasons, of which reason knows nothing." Or, less poetically, the Nobel laureate economist Friedrich Hayek noted that "if we stopped doing everything for which we do not know the reason, or for which we cannot provide a justification . . . we would probably soon be dead."[7]

How can we utilize this understanding of choice to our advantage? One place to start is in market choices. Montague and his colleagues scanned the brains of sixty-seven subjects inside Baylor's fMRI machine. Some of them received a sip of either Coke or Pepsi from a tube, while some of them were exposed to a picture of a distinctively labeled Coke or Pepsi can, or to no image at all. The subjects showed no preference for the colas offered with no label (in other words, they failed the classic "taste test"), but they overwhelmingly preferred any cola that was delivered along with the Coke brand. In analyzing the brain scans, Montague discovered that the Coke brand has a "flavor" in the *ventromedial prefrontal cortex*, a region essential for decision making. The Pepsi brand triggered no such brain response.

What this means is that certain brands change dopamine delivery to different parts of the brain. "These experiments show clearly that a cultural message, the brand image of Pepsi or Coke, has differential representation in people's nervous systems in such a way that this brand knowledge can be visualized in fMRI experiments, where its influence on choice can also be measured," Montague concluded.[8] (This is the same area, by the way, that was destroyed in the brain of the now notorious Phineas Gage, the nineteenth-century railroad worker who had an iron tamping rod accidentally blasted through his skull. Astonishingly, he survived, but he suffered lifelong social and emotional disabilities, most notably an inability to make decisions about even the most quotidian needs.) Brands stamp their power on our brains on a short-term basis through dopamine delivery and on a long-term basis by rewiring our neurons. And this process happens on a daily basis—it has been estimated that we are exposed to forty thousand commercials a year. By the age of eighteen months, children can recognize product logos. By age ten they know three hundred to four hundred brands by memory. By adulthood the number of recognized brands climbs into the thousands. "We can show that the idea of Coca-Cola activates structures in your midbrain that literally drive your behavior," Montague explains. "That is how ideas gain control over instinct."[9]

Just as Coke is a proxy for flavor, hunger a proxy for caloric need, lust a proxy for reproductive necessity, and guilt and pride proxies for immoral and moral behavior, so too can we market moral brands in order to reward and rewire brains to value and choose good ideas. This is what consciousness-raising is all about, and we now understand the neural wiring behind it. So in honor of the late economist Milton Friedman, author of the once radical but now mainstream book *Free to Choose*, which early in the development of my economic ideas rewired my brain's dopamine system to prefer the brand of free choice in markets, I propose that we begin our consciousness-raising for free societies by marketing this brand—the *Principle of Freedom*: *All people are free to think, believe, and act as they choose, as long as they do not infringe on the equal freedom of others.*

✦ ✦ ✦

In response to the Principle of Freedom, two challenges may be leveled: one is the *Overdeterminism Problem*, in which the many causal variables discovered by science appear to leave little room for genuine freedom and free choice; the second is the *Paternalism Problem*, in which critics of free markets and market capitalism could argue that people are too irrational and too determined to act wisely in their own interests and in the interests of society, and so politicians and lawmakers need to restrict our political and economic freedoms and make choices for us. I shall dispense with both of these challenges forthwith. After he published his book *Sociobiology* in the late 1970s, Edward O. Wilson was viciously attacked; when evolutionary psychology first made headway in the 1990s, there were equally vindictive assaults on its researchers. The genesis of these attacks was a fear that science would rob humans of our freedom and dignity. Nothing could be further from the truth. In fact, the best research to date shows that *at most* some human traits are 50 percent genetically or biologically determined.

Consider the gene that codes for the production of the brain neurotransmitter dopamine. Called D4DR, it is located on the short arm of the eleventh chromosome. When dopamine is released by certain neurons in the brain, it is picked up by other neurons that are receptive to its chemical structure, thereby establishing dopamine pathways that stimulate organisms to become more active and reward certain behaviors that then get repeated. Knocking out dopamine from a rat or a human, for example, causes them to go catatonic. Overstimulating dopamine causes frenetic behavior in rats and schizophrenic behavior in humans. We know about the D4DR gene because of the discovery by the geneticist Dean Hamer of its connection to risk-taking behavior. Most of us have four to seven copies of the D4DR gene on chromosome 11. Some people, however, have two or three long copies, while others have eight to eleven short copies. Longer, and fewer, copies of the D4DR gene translate into lower sensitivity to dopamine, which stimulates people to seek greater risks in order to artificially get their dopamine fix. Leaping off buildings, antennae, spans, or earth (from which the acronym BASE jumping comes) is one way to do it, although high-risk gambling in Las Vegas or on Wall Street may also do the trick. As a test of this hypothesis, Hamer first had subjects take a survey that measures desire to seek novelty and thrills. (Skydivers knock the ceiling off this survey.) He then took a sample of their DNA from chromosome 11 and discovered that people who score high on the risk-taking survey had fewer copies than normal of the D4DR gene.

This sounds deterministic in headline form ("Scientists Discover Risk-Taking Gene"), but in fact Hamer claims that this finding enables us to explain only 4 percent of novelty-seeking and risk-taking behavior based on the D4DR gene sequences alone. Recall that in a previous chapter we discussed correlation and what it means to square the r to give us a percentage of the variance in any given trait that can be accounted for by the variable under question. Height and weight give an r of 0.70, or $r^2 = 0.49$, so we can say that 49 percent of one's weight is accounted for by one's height, which means that fully half of your weight is under the control of environmental conditions, such as diet and exercise. Think about that finding in the context of the Overdeterminism Problem. You won't find a trait much more biologically and genetically determined than the relationship of height and weight, and yet even here you get to control half of the variance yourself by freely choosing to manipulate your environment and lifestyle choices.

So when neuroscientists claim to have discovered a brain module or a neural circuit associated with X, as important and interesting as such findings are, it is anything but deterministic in its implications. And dopamine is a perfect example for our purposes because it is one of the most exciting and potentially useful discoveries ever made by neuroscientists, but if the genetic architecture determining the output of dopamine in the brain accounts for only 4 percent of the variance among people on some given trait—such as risk-taking behavior—how can that possibly lead us to conclude that we are determined by the biology of our brains?[10]

Of course, one might rejoin that this is just one of numerous brain chemicals that in cocktail form intoxicate us into taking actions we might not otherwise choose. Given the suite of findings from behavioral economics and neuroeconomics that have demonstrated just how unconscious and irrational our choices are, it may seem reasonable to call into question just how much freedom we have when we feel free to choose. And we haven't even considered the most dramatic research on the neuroscience of free will—that of Benjamin Libet in his now famous 1985 experiment, since corroborated in many labs. Libet took EEG readings of the brain's activities in order to determine the precise moment at which we become aware of our intention to perform an action, such as raising a finger. Subjects sat before a screen in which a dot was moving about a circle like the second hand on a clock face. They were asked to do two things: (1) note the position of the dot on the screen when "he/she was first aware of the wish or urge to act," and (2) press a button that also recorded the position of the dot on the screen. The difference between (1) and (2) was two hundred milliseconds. That is, two-tenths of a second elapsed between thinking about pressing the button and actually pressing it.

But that's not the disturbing part. The EEG recordings for each trial revealed that the brain activity involved in the initiation of the action was primarily centered in the secondary motor cortex, and that part of the brain became active three hundred milliseconds before subjects reported their first awareness of a conscious decision to act. That is, the awareness of our intention to do something trails the initial wave of brain activity associated with that action (what is called the EEG "readiness potential") by about three hundred milliseconds. That is, three-tenths of a second elapsed between the brain's making a choice and our awareness of the choice. Add to that three-tenths of a second the other two-tenths of a second to act on the choice, and it means that a full half a second passes between our brain's intention to do something and our awareness of the actual doing of the act. Because the neural activity that precedes the intention to act is inaccessible to our consciousness mind, we experience a sense of free will. But it is an illusion, caused by the fact that we cannot identify the cause of the awareness of our intention to act. Because the action consistently follows the intention, we feel as if we freely willed the act, when in reality both the awareness of intention and the overt action in response to the intention are caused by prior neural events of which we are unaware.[11]

Such findings imply that our "free choices" are really nothing more than the equivalent of the magician who offers a volunteer from the audience a chance to "pick a card, any card," knowing full well that he is employing some version of a "force" that makes the volunteer's choice anything but free. This is a very serious problem for economists, politicians, and social theorists, because to live in a civil society we need to hold people accountable for their actions, which means that we must assume that they make free choices. If we are fully determined by a combination of our genetics and our environment, then how can we justify punishing someone for an illegal or immoral act? Here are several solutions to this conundrum that I believe maintain the integrity of both science and society:[12]

Free will as a useful fiction. I feel "as if" I have free will, even though I know we live in a determined universe. You do the same. Since no one has satisfactorily solved the problem of free will and determinism in four thousand years of philosophical thought and five hundred years of scientific research, the problem may be an insoluble one. (The insolubility, in fact, may be due to nothing more than the limitations of language. Depending on how "free will" and "determinism" are defined, it may simply be impossible to square the circle.) So why not act as if you do have free will, thereby gaining the emotional gratification that comes with this useful fiction, along with the social benefits that accrue by holding people accountable for their actions?

Free will as a fuzzy fraction of determinism. Instead of thinking of concepts like "free will" and "determinism" as reified things—Platonic types that exist as unchanging entities—reconfigure them in the language of *fuzzy logic*, where we assign a fractional probability to something. Just as the sky can be 0.1 blue at dawn and 0.9 blue at midday, so too can behavior be assigned a fuzzy probability; for example, perhaps someone's behavior could be scaled somewhere between a 0.1 and a 0.9 in evilness, or between a 0.1 and a 0.9 in how much the evil behavior was freely chosen or shaped by other forces. The law already makes such fractional distinctions for homicides that range between, say, 0.1 for an accidental shooting, 0.5 for a self-defense shooting, and 0.9 for a premeditated shooting. A behavior that ranges in causal influence between 0.1 and 0.9 is still not an absolute 0 or 1, and so moral culpability is sustained through the fractional level of free will that remains. Even if free will is diminished, it is not extinguished.

Free will as causal uncertainty. In science, the causal-net theory of determinism holds that human behavior is no less caused than other physical or biological phenomena, but that it is more difficult to understand and predict because of the number of causal elements in the net encompassing our behavior. The human and social world in which we live is extremely complex, interactive, and loaded with autocatalytic feedback loops that drive systems into states of chaotic behavior. Since no cause or set of causes we consider as the determiners of human action can be complete, practically speaking we can treat them as *conditioning causes*, not determining ones. Although our genes, environments, and historical pathways on some level do determine our actions, every individual set of genes is unique, each environmental setting is distinctive, and every historical pathway that each of us has taken belongs to us alone. In this sense, each of us is unique and different from all others, the product of matchless genes, environments, and historical pathways that are so complex and so entwined that no one could possibly know all of the causal variables for themselves or anyone else. Free will arises out of this ignorance of causes.

Free will as an evolutionary by-product. Because of our uniquely huge brain and exceptionally large prefrontal cortex, we are self-aware and aware that others are self-aware. We have a Theory of Mind that allows us to place ourselves inside others' minds, and we know that others have the same capacity. We can reason using symbolic language that also allows us to communicate and reason with others. We are moral animals with an evolved sense of right and wrong and a natural inclination to be both cooperative and competitive, altruistic and selfish, good and evil. Free will emerges from the fact that we can weigh the consequences of the many courses of action available to us, then make choices and act on them, and also that we are aware that we (and others) make such choices. From these evolved capacities we can and do hold ourselves accountable for our actions just as we can and do hold others accountable for their actions.

Free will as an emergent property of the brain. The mind is an emergent property of billions of individual neurons, each one of which is connected to thousands of other neurons that together produce trillions of potential neuronal states. As the individual grows and develops into adulthood, the interconnections grow and develop according to individual life experiences. Although we share a common evolutionary ancestry that generated a universal neural architecture, the brain changes in response to the environment. This sets up another self-generating feedback loop in which new experiences stimulate neurons to grow new synaptic connections. Those new connections are distinctive to every individual mind, which then responds to the environment in a unique manner, producing a behavioral repertoire of responses that is unmatched by anyone else's. Since no life paths are the same, the trillions of possible permutations of neuronal connections in each brain mean that every human mind is unique. Out of the higher order emergent property of this uniqueness, in conjunction with the previous freedom factors, comes free will.

Free will as a product of neural computation. In this book I have argued that states of mind such as emotions are efficient proxies for computations made by evolution over the Paleolithic eons so that we don't have to make the calculations ourselves. In this sense, free will may be a proxy for choice computation. Making choices is a real neural process that involves selecting behaviors that have survival consequences, such as predator avoidance, food preparation, mate selection, friendship bonding, social status seeking, and the like. Thus, making choices that lead to behaviors that result in actual consequences for survival and reproduction in our evolutionary history would have led to the evolution of brain mechanisms that give a sense of free will, a feeling of freedom, an emotion of volition. With the complex brains that we possess, living in a world with so many options from which to choose, our brains have built into them a choice-making module that, whether truly free or truly determined, nonetheless makes us feel free.

✦ ✦ ✦

The *Paternalism Problem* arises directly from the research on Subjective Well-Being and happiness. If one goal of society is to create the greatest happiness for the greatest number, to what extent should public and private institutions attempt to establish policy to exacerbate sins and enhance virtues? Since research shows that money does not make people happier unless they are below the poverty line, does this justify a form of happiness welfare to ensure that everyone is at least out of poverty? Likewise, science reveals that, on average, married people are happier than divorced and single people, so should government incentives such as tax breaks for marriage be increased to encourage more participation in this social institution? Since people with meaningful work tend to be happier than the unemployed, does this justify government make-work programs? Religious people are slightly happier than nonreligious people, so should the state create religious tax incentives?

According to the London School of Economics economist Richard Layard, such paternalisms are justified by the scientific research.[13] Using the findings from behavioral economics, neuroeconomics, and other related sciences that inform us about what makes people happy and why, Layard concludes that governments should get more involved in directing the personal lives of their citizens. Since we know that smoking and drinking are unhealthful, we should increase taxes on such behavior by passing more so-called "sin taxes." Because a large disparity in wealth between the rich and the poor makes people in lower income brackets feel that they can never measure up, this would justify an increase in taxes on the rich in order to redistribute wealth and discourage a rat-race mentality. Research reveals that we get great satisfaction from helping the poor and the mentally ill, so we should establish incentives to encourage more people to do so. Unemployment dramatically decreases happiness, so government must take steps to eliminate high unemployment even if it means setting up government work programs just for this purpose. Psychologists have proved that excessive commercial advertising to children increases their desire for material things that are really not necessary to experience a fulfilling childhood, so such advertising practices should be banned. Furthermore, Layard thinks that we need to enforce more family-friendly practices at work, subsidize social activities that encourage community spirit, and include in public school curricula a K–12 yearly course in moral education that would cover principles of morality, the practice of empathy, the importance of serving others, the study of role models, the control of emotions, parenting, mental illness, and how to be a good citizen. "This means that public policy should be judged by how it increases human happiness and reduces human misery," Layard suggests, citing the research that shows that "extra income increases happiness less and less as people get richer." From this he concludes, "This was the traditional argument for redistributive taxation, and modern happiness research confirms it."[14] Brave new world.

The paternalism conclusion is not even wrong. *Happiness. Freedom. Government.* Pick two. If you want happiness by government fiat, there goes your freedom. Enforced policies to encourage happiness must inescapably lead to a decrease in freedom. This is true by definition— enforced means you are forced; that is, you have no choice. If you want happiness and freedom, you have to minimize government interference. As Jefferson warned, "A government big enough to give you everything you want, is strong enough to take everything you have."

The Austrian economist Ludwig von Mises, the *spiritus rector* of free market economics, demonstrated why the wrong mixture of government and freedom leads to tyranny and unhappiness. He learned the lesson first in his personal life. Mises was born in 1881 within the then powerful

Austro-Hungarian Empire. He studied law and economics at the University of Vienna under Friedrich von Wieser and Eugen von Böhm-Bawerk, both followers of Carl Menger, the founder of the Austrian School of Economics. After serving as an artillery officer on the Russian front in World War I, Mises earned international recognition for his first major book, *Socialism*, in which he demonstrated why planned economies cannot work without a free market pricing system. He continued working on specific problems in economic theory, until his life was disrupted in March 1938 when Hitler marched into Vienna and Mises marched out to the United States. There Mises began his long and lonely struggle against economic and political tyranny, a lone advocate of economic freedom in an increasingly socialistic society. The problem, Mises argued as he expanded his theory to encompass not just the economic sphere but the political as well, is that government interventionism in one area leads to additional interventionism in other areas. In his 1949 magnum opus, *Human Action*, Mises posed this problem: if you allow governments to paternalistically intervene to protect individuals from dangerous drugs, what about dangerous ideas?

> Opium and morphine are certainly dangerous and habit forming drugs. But once a principle is admitted that it is the duty of government to protect the individual against his own foolishness, no serious objections can be advanced against further encroachments. A good case could be made out in favor of the prohibition of alcohol and nicotine. And why limit the government's benevolent providence to the protection of the individual's body only? Is not the harm a man can inflict on his mind and soul even more disastrous than bodily evils? Why not prevent him from reading bad books and seeing bad plays, from looking at bad paintings and statues and from hearing bad music?[15]

Mihaly Csikszentmihalyi, the psychologist who discovered flow (whose work we discussed in chapter 8), gave me an example of too much state paternalism in people's personal lives, which he recalled was imposed in his home country while it was under Communist control: "Back in Hungary they figured that popular products like Beatles records should be taxed in addition to the normal tax. They called this a *garbage tax*, because the government thought of such products as garbage. They then applied that tax money to funding what they considered to be more culturally worthwhile projects, such as the symphony, the ballet, etc. It's very paternalistic, and yet they claimed they are on the side of culture and tradition. But it didn't work."[16]

If research shows that the existence of wealthy neighbors puts me on a hedonic treadmill that I can never satisfy, and therefore policy is legislated forcing my neighbors to redistribute some of their wealth to me and others less fortunate, this will not increase my happiness, because I did not earn the bonus, and it will not increase my neighbor's happiness, because they did not voluntarily donate their wealth to what they deemed a good cause. Scientific research shows that economic self-reliance makes people happier than economic dependency, and studies show that people are happier, healthier, and more generous when they voluntarily donate their money to causes they deem worthy instead of having their money confiscated from them and given to causes that they would not otherwise have chosen to support. Proof for this claim can be found in two sets of data: (1) studies on national charitable giving, and (2) studies on international happiness and freedom.

1. *National Charitable Giving.* Research on the difference between forced and voluntary giving reveals a counterintuitive finding on the differences between the political left and right. If we are going to base political policy on scientific data, then what are we to make of the research reported by Syracuse University professor of public administration Arthur C. Brooks in his revealing book *Who Really Cares?* When it comes to charitable giving and volunteering, numerous quantitative measures debunk the myth of "bleeding-heart liberals" and "heartless conservatives." The opposite, in fact, appears to be true. Conservatives donate 30 percent more money than liberals (even when controlled for income), give more blood, and log more volunteer hours. And it isn't because they have expendable income that conservatives are more generous. The working poor give a substantially higher percentage of their incomes to charity than any other income group, and three times more than those on public assistance of comparable income. In other words, poverty is not a barrier to charity, but welfare is. One explanation for these findings is that people who are skeptical of big government give

more than those who believe that the government should take care of the poor. "For many people," Brooks explains, "the desire to donate other people's money displaces the act of giving one's own."

On that front, we don't need science to tell us what we already know. The French economist Frédéric Bastiat, whom I have quoted on several occasions already, wrote in the early nineteenth century, "Government is the great fiction, through which everybody endeavors to live at the expense of everybody else."[17] Later that century George Bernard Shaw noted with sarcasm, "A government which robs Peter to pay Paul can always depend on the support of Paul."[18] Or as G. Gordon Liddy put it in his straight-faced style, "A liberal is someone who feels a great debt to his fellow man . . . which debt he proposes to pay off with your money."[19] Finally, as the political humorist P. J. O'Rourke observed, "Giving money and power to government is like giving whiskey and car keys to teenage boys."[20]

Liberals feel that they already donated to the poor through their taxes, whereas conservatives believe that it is *their* duty, not the government's, to assist those in need. Let's think about these findings in the context of evolutionary economics. Our entire evolutionary history was played out in tiny social groups of highly related family and tightly bonded relationships where mutual aid and cooperative support were vital to ensure the survival of the group members and the group itself. We have already seen how religion is a proxy for tight social bonding among members of a community, so it should not surprise us to learn that religious people are four times more generous than secularists to all charities, 10 percent more munificent to nonreligious charities, and 57 percent more likely than a secularist to help a homeless person. Those raised in intact and religious families are more charitable than those who were not. And the adaptive survival benefits of giving are real: in terms of societal health, charitable givers are 43 percent more likely to say they are "very happy" than nongivers, and 25 percent more likely than nongivers to say their health is "excellent" or "very good."[21]

By the paternalistic line of reasoning, then, and following these data where they lead to public policy, governments should give tax breaks to conservatives, the wealthy, the working poor, and the religious in order to reward their prosocial behavior and encourage more giving. All tax-and-spend liberals in favor of this policy are invited to my home for beer and burgers.

2. *Studies on International Happiness and Freedom*. International research shows that an increase in personal autonomy and self-control leads to greater happiness, and that people tend to be happier in societies with greater levels of individual autonomy and freedom compared to those in more totalitarian and collectivist regimes. The social scientist Ruut Veenhoven, from Erasmus University in Rotterdam, for example, conducted a comprehensive survey on happiness with life as a function of three social conditions: individualism, opportunity to choose, and capability to choose. "The data show a clear positive relationship," Veenhoven concludes; "the more individualized the nation, the more citizens enjoy their life." Further, he found no "pattern of diminishing returns," meaning that "individualization has not yet passed its optimum." In other words, greater levels of individual freedom and autonomy could lead to even greater levels of happiness.[22]

The rub is in finding the right balance between individual freedom and autonomy and collective fairness and justice. In the small bands of our Stone Age ancestors, the social ties of genetic relatedness and the social glue of reciprocal exchange served as the natural cohesive to hold people together without a lot of external constraints on freedom and autonomy, but with the rise of chiefdoms and states, artificial institutions were needed to enforce the rules of cooperation and conflict resolution. But how much external governance do we need?

✦ ✦ ✦

How can we paternalistically encourage behaviors that lead to greater mental health and happiness and not at the same time decrease freedom and choice, thereby obviating the original purpose of the paternalistic policy? The answer may be found in programs that allow the maximum freedom of choice while providing incentives for options that encourage healthier and happier living. An example of just such a program is called *libertarian paternalism*, developed by the University of Chicago economists Cass Sunstein and Richard Thaler. They swear the term is not an oxymoron.[23] Sometimes called "soft paternalism"—to contrast it with the "hard paternalism" of strong state interventionism—libertarian paternalism preserves freedom of choice while implementing what we have learned from behavioral economics and neuroeconomics about people's needs, wants, and irrationalities. Sunstein and Thaler propose that we encourage the establishment of policies and institutions that nudge people in the direction of doing what is good for them, as informed by science and freely chosen by voluntary consent. Reflecting the language of behavioral economics, the authors write:

> Often people's preferences are ill-formed, and their choices will inevitably be influenced by default rules, framing effects, and starting points. In these circumstances, a form of paternalism cannot be avoided. Equipped with an understanding of behavioral findings of bounded rationality and bounded self-control, libertarian paternalists should attempt to steer people's choices in welfare-promoting directions without eliminating freedom of choice.[24]

We know, for example, that when people are assessing a medical procedure that includes significant risk, how the options are presented significantly influences the choice that is made—patients are far more likely to agree to a risky procedure when they are told, "Of those who have this procedure, 90 percent are alive after five years," than if they are told, "Of those who have this procedure, 10 percent are dead after five years."[25] So there is a difference between feeling lucky to be alive and feeling lucky to be not dead, and that difference will be reflected in how the question is asked. Since the question has to be worded in some manner, that phrasing should be as well informed by science as possible.

Another constraint on choice is the nature and origins of the options from which to choose. As a simple example, if you are the owner of a restaurant, you have to design a menu that you think will appeal to customers. There must be some arrangement of the items on the menu, and since *you* are going to make that arrangement, you can fill the menu with nothing but healthful foods that taste bland, nothing but unhealthful foods that taste great, or some admixture thereof. Whatever arrangement you make, you have just limited the choices of your customers. If, say, in addition to wanting to make a modest profit in order to keep your business going, you also have a social conscience and would like to help society by encouraging people to consume more healthful diets, you could paternalistically offer people nothing but healthful choices. Ideally, those food items would also be appetizing in order to meet your first need of staying in business, but once that need is met, you can move up the moral hierarchy by prodding your customers toward better dietary habits.

In fact, in March 2007, the restaurant chain T.G.I. Friday's implemented something very much like this with their new menu they call *Right Portion, Right Price*, in which they offer customers smaller-portioned meals at a lower price. The CEO, Richard Snead, offered an explanation that perfectly balances the tension between corporate profit and social conscience: "This is a category issue stemming from consumer demand. The category needs to listen. This is a significant part of Friday's overall goal of personalizing and customizing the guest experience. No matter what your lifestyle choice, you don't have to sacrifice taste. Smaller portions at smaller prices meet all lifestyle choices."[26]

Where there is a binary choice that must be made, an effective strategy is to switch the default option from electing to participate (opt in) to choosing not to participate (opt out). Since people have to make a choice one way or the other—either to participate in the program or not—why not structure the choice that will lead to the greatest social good? Sunstein and Thaler call this *libertarian benevolence*, a corollary to libertarian paternalism. For example, in the United States we have an opt-in policy for organ donation, where you have to actively punch a little tab on your driver's license in order to consent to having your organs removed in the event of your death. In countries where organ donation is based on an opt-out policy—such as Austria, Belgium, Denmark, Finland, France, Italy, Luxembourg, Norway, Singapore, Slovenia, and Spain—your organs will be harvested in the event of your death unless you actively choose not to allow it. In countries with an opt-out policy, organ donation participation is on the order of over 90 percent, compared to under 20 percent for countries employing the opt-in strategy.[27] Here we have preserved freedom of choice, but by altering the default options we have made a significant difference in the social outcome.

Corporations can employ a similar opt-out system. The next step for T.G.I. Friday's, for example, would be for them to automatically offer customers the *Right Portion, Right Price* menu and force them to request the menu with the larger portions and higher prices if they want to switch.[28] It's a time-tested system in corporations already, as when companies automatically enroll employees in pension plans. Instead of asking workers if they would like to participate in the company 401(k) plan, for example, the company just automatically enrolls them unless they actively choose to opt out. Companies that have implemented this system have experienced increases in pension plan enrollments as high as 40 percent. A related program automatically withholds money from people's paychecks, and then gives them a choice of several financial instruments in which to invest that money. This is a libertarian policy because people still have a choice, but it is paternalistic because extensive research shows that most people are clueless when it comes to investing, whereas companies at least consult with investment experts so that all of the choices offered to employees are sound and reliable investment vehicles.

Libertarian paternalism makes a deeper assumption about our nature—that at our core we are moral beings with a deep and intuitive sense about what is right and wrong, and that most of the time most people in most circumstances choose to do the right thing. Under that principle, the default option should be to grant people the libertarian ideal of maximum freedom, while using the best science available to inform the policies that give structure to the minimum number of restrictions on our freedoms. Let's opt for more freedom and add back restrictions on freedom only where absolutely necessary and with great reluctance.

TO OPEN THE WORLD

In his magnum opus on the power of free minds and free markets, *Human Action*, the Austrian economist Ludwig von Mises observed: "The truth is that capitalism has not only multiplied population figures but at the same time improved the people's standard of living in an unprecedented way. Neither economic thinking nor historical experience suggest that any other social system could be as beneficial to the masses as capitalism. The results speak for themselves. The market economy needs no apologists and propagandists. It can apply to itself the words of Sir Christopher Wren's epitaph in St. Paul's: *Si monumentum requires, circumspice.*" If you seek his monument, look around.

Capitalism may not need apologists and propagandists, but it does need a scientific foundation grounded in psychology and evolution, which I have attempted to give it in this book. Now I would also like to look into the future.

For many years, I have been involved in a Seattle-based organization called Foundation for the Future, created by the aerospace entrepreneur and philanthropist Walter Kistler, in which a group of scientists and scholars from various fields meet once a year to discuss what life will be like in the year 3000, among other lofty topics.[1] It is a delightfully stimulating way to spend a weekend, but I do not for a moment think that any of us has any idea what we are talking about when we consider life a thousand years from now. If most experts on the Soviet Union in the mid-1980s had no idea that the empire would collapse by the end of the decade, and if most computer scientists in the early 1980s were largely clueless to the forthcoming rise of the World Wide Web, how on earth can anyone possibly fathom what changes will be wrought a hundred decades from now?[2]

The problem with envisioning long-term economic and political change is that we have been entrenched in political states with top-down-directed economies for so many millennia that it is nigh impossible to imagine how human relations could peacefully prosper in any social system other than the one to which we have grown accustomed. By the logic of the *status-quo bias*, nature has endowed us to hold dear what is ours and leads us to opt for whatever it is we are used to having. Still, history's long clock and evolution's deep time afford us the opportunity to pull back and see the bigger picture of what all this research on markets, minds, and morals means for the eventual liberation of humanity.

◆ ◆ ◆

We began this book with the really hard problem of explaining the great leap forward from a hunter-gathering economy to a consumer-trading economy. Employing scientific tools and data from complexity theory, evolutionary biology, behavioral psychology, and neuroscience, we saw that the economy is a complex adaptive system that changed and adapted to circumstances as it evolved out of a much simpler system, and that the first ninety thousand years of our tenure as hunter-gatherers created a psychology that has often led us to behave irrationally in the last ten thousand years of the great leap forward.

No one has stimulated me to further understand the dynamics of this economic shift more than Jared Diamond, who is just about the most interesting interdisciplinary polymath I have ever had the pleasure to know. A slightly built man with a resonant baritone voice and seamless delivery style that leaves audiences in anticipation of the next insight, in informal settings Jared's modest demeanor belies the depth of insight and breadth of knowledge needed to answer one of history's grandest mysteries, which I believe has direct bearing on the overall goal of this book of explaining how hunter-gatherers became consumer-traders. Here is the mystery:

Sometime between about thirty-five thousand and thirteen thousand years ago, the tool kits of early humans became much more complex and varied than they had ever been before. Suddenly, sewing led to clothing to cover our now nearly hairless and naked bodies. The first homes were built of bones and wood and animal skins to shelter us from the climate. Sophisticated representational art was being created deep inside caves, and symbolic communication led to the development of complex spoken languages. With these and many other changes, anatomically modern humans began to wrap themselves in a blanket of technology, forever altering natural selection and taking evolution into their own hands. These prehistoric humans soon spread to nearly every region of the globe, and everyone everywhere lived in a condition of hunting, fishing, and gathering. Some were nomadic, while others stayed in one place. As small bands grew into larger tribes, possessions became valuable, rules of conduct grew more complex, and population numbers crept steadily upward.[3]

Then, at the end of the last Ice Age, roughly thirteen thousand years ago, population numbers in several places around the globe exploded in size. Hunting, fishing, and gathering did not produce enough calories to support these larger populations, and the need for sustenance led inexorably to farming and the Neolithic Revolution. There is considerable debate among archaeologists and anthropologists about how and why this revolution in food production came about, how long it took to make the transition, and whether it was a continuous (evolutionary) or discontinuous (revolutionary) change. For our purposes here, the domestication of large mammals and edible grains generated the necessary calories to support ever-increasing populations, which led to additional physical and social technologies that facilitated even larger populations, and so on in the now familiar autocatalytic feedback loop that drives such self-organized emergent systems.[4]

Then something odd happened. Between thirteen thousand years ago and today, there was a noticeable difference in the rates of development between civilizations around the globe, with some accelerating dramatically toward modernity while others remained mired in the Paleolithic mud. To put it in the form of a question: why is it during the last five hundred of those thirteen thousand years that Europeans conquered and colonized the Americas and Australia, rather than Native Americans and Australian Aborigines conquering and colonizing Europe? Rejecting antiquated explanations rooted in inherited racial differences that facilitated some races (white) and impeded other races (black), Diamond posits that the differences in rates of development between civilizations around the world over those thirteen millennia were primarily the result of geographical differences in the availability of domesticable grains and animals, which in some areas (but not others) enabled the development of farming, large populations, division of labor, non-food-producing specialists, metallurgy, writing, military, government bureaucracies, and the other necessary components that ultimately gave rise to modern civilization.[5]

The indigenous wild grains that could be domesticated, for example, were few in number and located only in certain regions of the globe—those regions that saw the rise of the first civilizations. Australian Aborigines, to cite another example, could not strap a plow to or mount the back of a kangaroo, as Europeans did the ox and horse. Additional factors include the east-west-oriented axis of the Eurasian continent, which lent itself to the diffusion of domesticated grains and animals as well as of knowledge and ideas, so that Europe was able to benefit much earlier from the domestication process. The north-south-oriented axis of the Americas, Africa, and the Asia-Malaysia-Australia corridor, by contrast, did not lend itself to such fluid transportation, and thus those areas not already well suited biogeographically for farming could not even benefit from trade and diffusion of such foods and farming technologies. Finally, through constant interactions with domesticated animals and other peoples, Eurasians developed immunities to numerous diseases that, when brought by them in the form of germs (along with their guns and steel) to Australia, Oceania, and the Americas, produced genocides on a hitherto unseen scale.

✦ ✦ ✦

How and when we made the transition from small bands and tribes to large chiefdoms and states was determined in part by the carrying capacity of the environment and the population size of the groups that in turn determined the structure of societies and the forms of exchange, trade, and coexistence with other groups.[6] The concomitant leap in food production and population was accompanied by a shift from bands and tribes to chiefdoms and states, and the development of appropriate social organizations and technologies. People began to live in semipermanent and then permanent settlements, which led to land ownership and private property, and to surplus foods, tools, and other products that formed the basis of nascent trading economies. This led naturally to a division of labor in both economic and social spheres. Full-time artisans, craftsmen, and scribes worked within a social structure organized and run by full-time politicians, statesmen, and bureaucrats. Organized religion came of age to fill many roles, not the least of which was the justification of power for the ruling elite. The intertwining of politics and religion has been found in nearly every chiefdom and state society around the world, including the Middle East, Near East, Far East, North and South America, and the Polynesian Pacific islands, in which the chief, pharaoh, king, queen, monarch, emperor, sovereign, or ruler of whatever title claimed a relationship to God or the gods, who purportedly invested them with the power to act on behalf of the deity. States developed into bona fide civilizations, small sects evolved into world religions, and barter markets emerged into full-fledged economies.

With the rise of chiefdoms, states, and empires, it was no longer possible to separate politics from economics. Although the natural condition of hunter-gatherer bands and tribes is one of egalitarianism, the equal redistribution of economic wealth has never been realized in larger societies. Moreover, without the proper social institutions to enable and enforce fair and free exchange between groups, violence and war often erupt. Here too we find an evolutionary economic explanation. One of the prime triggers of between-group violence is competition for scarce resources. There are rarely enough means to support all individuals in all groups. Even if, at some given time, there were, such a condition could only be a temporary one, because populations naturally tend to increase to the carrying capacity of the environment. Once that capacity is exceeded, the demand for those resources will exceed the supply. Such was the condition throughout most of history for most peoples in most places. The formula is straightforward: *population abundance plus resource scarcity equals conflict*. Thus, one way to attenuate between-group violence is to increase the supply of resources to meet the demands of those in need of them.

◆ ◆ ◆

The psychology behind defusing intergroup aggression involves turning potentially dangerous total strangers into prospectively helpful honorary friends. This process is enabled through the creation of social institutions that encourage, enable, and enforce positive social interactions that lead to trust. One of the most powerful of these forms of interactions is trade, the effects of which I want to elevate into a principle based on an observation by the nineteenth-century French economist Frédéric Bastiat: "Where goods do not cross frontiers, armies will."[7]

Bastiat's Principle not only helps us understand how hunter-gatherers made the transition to consumer-traders, it also illuminates one of the primary causes of conflict; its corollary elucidates one of the principal steps toward conflict reduction. If Bastiat's Principle holds that *where goods do not cross frontiers, armies will*, then its corollary dictates that *where goods do cross frontiers, armies will not*. This is a principle, not a law, since there are exceptions both historically and today. Trade will not prevent war, but it attenuates its likelihood. Thinking in terms of probabilities instead of absolutes—fuzzy logic's range of fractional possibilities versus Aristotelian logic's A or non-A categories—trade between groups increases the probability that peaceful and stable relations will continue and decreases the probability that instabilities and conflicts will erupt.

Let's return to where we began this book with the Yanomamö hunter-gatherers and how they evolved into the Manhattan consumer-traders. When missionaries first began working with the Yanomamö, they discovered that if they provided the native peoples with tools for the procurement and production of food and other resources, the amount of Yanomamö intervillage fighting was greatly reduced. The great Yanomamö ethnographer Napoleon Chagnon—who originally gave the Yanomamö their "fierce people" moniker—discovered that the Yanomamö are sophisticated traders as well as ferocious warriors, because trade creates political alliances. Following the political dictum "The enemy of my enemy is my friend," Yanomamö intervillage trade and reciprocal food exchanges serve as a powerful social glue in the creation of political alliances. Village A cannot go to Village B and announce that they are worried about being conquered by the more powerful Village C, since that would reveal their own weakness. Instead, Village A forms an alliance with Village B through trade and reciprocal feasting, and as a result they not only gain military protection, but also encourage intervillage peace. As a by-product of this politically motivated economic exchange, even though each Yanomamö band could produce all the SKUs it needs for survival, they often set up a division of labor and a system of trade. The unintended consequence is an increase in both wealth and SKUs. The Yanomamö trade not because they are innate altruists or nascent capitalists, but because they want to form political alliances. "Without these frequent contacts with neighbors," Chagnon explains, "alliances would be much slower in formation and would be even more unstable once formed. A prerequisite to stable alliance is repetitive visiting and feasting, and the trading mechanism serves to bring about these visits."[8] *Where goods cross Yanomamö frontiers, Yanomamö armies do not.*

To make the point with a second example, in his study of two Australian Aboriginal "bush" people from the Western Australian desert, the Walmadjeri and Gugadja, the anthropologist Ronald Berndt notes that desert economics begins with the close ties between kin, who depend on knowing what to expect from others and what is expected of them. "The horde or band is not just a group made up of nucleated family units. . . . It is a cooperative unit, with each member caught up in an intimate network of responsibilities and obligations, depending on others as others depend on him." Because the Western Australian desert is such a brutally harsh environment, many of their religious and magical rituals are steeped in the physical necessities of life—most notably water—and trade lines established between groups are typically oriented along paths in which such commodities can be found. With such limited resources, the potential for conflict is high, but the consequences of engaging in conflict are also grave, and so the desert Australian aborigines have developed a system of trade intimately tied to their religion and the environment in order to foster goodwill between groups. "When large ceremonies and rituals are held, some of the participants come from places a great distance apart; they provide, therefore, an ideal opportunity for bartering," Berndt explains. "Trade takes place within the context of ritual and often is not seen as being something separate."[9] *Where goods cross the frontiers between Australian aboriginal groups, their armies do not.*

In a third example, Jared Diamond notes how caution and distrust is the norm between strangers among the hunter-gatherers of New Guinea he has lived among and studied for some three decades. Diamond's experiences there afforded him the opportunity to see firsthand how deeply dependent is trust on personal and social relations. Social obligations, Jared notes, depend on human relationships. "Because a band or tribe contains only a few dozen or a few hundred individuals respectively, everyone in the band or tribe knows everyone else and their relationships. One owes different obligations to different blood relatives, to relatives by marriage, to members of one's own clan, and to fellow villagers belonging to a different clan." Lacking the social institutions consumer-traders employ for conflict resolution, disputes among hunter-gatherers are directly resolved because within these small bands everyone is related to one another or knows one another, and members of the band are distinctly different from nonmembers on all levels, generating within-group amity and between-group enmity. Tribalism rules the day. "Should you happen to meet an unfamiliar person in the forest, of course you try to kill him or else to run away," Diamond explained. "Our modern custom of just saying hello and starting a friendly chat would be suicidal."[10] And yet something happened in the 1960s to bring about more peaceful interactions. Initially, peace was imposed upon the native New Guineans by fiat from the Western colonial government that ruled over the territory, but officials then ensured continued peace by providing goods that the people needed, as well as the technologies to enable them to continue producing more resources on their own. In less than one generation, New Guinean hunter-gatherers who were fighting each other with stone tools were suddenly New Guinean consumer-traders operating computers, flying planes, and running their own small businesses.[11] *Where goods crossed New Guinea frontiers, New Guinea armies did not.*

To reiterate, although trade is not a surefire prophylactic for between-group conflict, it is an integral component to establishing trust between strangers that lessens the potential volatility that naturally exists whenever groups come into contact with one another, especially over the allocation of scarce resources.[12] Moreover, since I have tightly linked market capitalism with liberal democracy in this book, I should note that there is a well-documented correlation between liberal democracy and peace—the more a nation embraces liberal democracy, the less likely it is to go to war, especially against another liberal democracy. The political scientist Rudolf J. Rummel has researched this relationship thoroughly, showing in one study, for example, that of the 371 international wars that occurred between 1816 and 2005 in which at least a thousand people were killed, there were 205 wars between nondemocratic nations, 166 wars between democratic and nondemocratic nations, and no wars between democratic nations. From this and many other historical data sets, Rummel draws five conclusions: "First, well established democracies do not make war on and rarely commit lesser violence against each other. Second, the more two nations are democratic, the less likely is war or lesser violence between them. Third, the more a nation is democratic, the less severe its overall foreign violence. Fourth, in general the more democratic a nation, the less likely it will have domestic collective violence. Finally, in general the more democratic a nation, the less its democide [the murder of its own citizens]."[13]

Conclusion: *Power kills; democracy saves.* Solution: *Spread democracy.* From the economic data and theory presented in this book, I would add the following. Conclusion: *Trade leads to peace and prosperity.* Solution: *Spread trade.*

This is oversimplified, of course, and an epilogue is no place to launch a discussion of the massive literature on the history, politics, and economics of war.[14] My larger point here is this: just as I have argued that morality in the form of moral emotions evolved long before religion and politics developed historically,[15] I am claiming that trade evolved long before the state developed economic institutions of trade, and thus we evolved moral emotions linking trade and trust, and this link is directly related to intergroup war and peace. It appears, for example, that trade between human groups dates back at least two hundred thousand years, as archaeologists have found stone tools and other artifacts such as seashells, flint, mammoth ivory, and beads hundreds of miles from where they were manufactured.[16] A more recent example can be found in Native Americans, who were active traders when European explorers and colonists arrived—the archaeologist Shepard Krech makes the point that the reason Europeans were so readily able to trade with the American natives (swapping beads for pelts, for example) was that the indigenous Americans were already well accustomed to trading among themselves.[17]

Trade breaks down the natural animosities between strangers while simultaneously elevating trust. Recall the research reviewed in chapter 9 on what happens when strangers trade. Dopamine (the lust liquor that is also related to addictive behaviors) is released, which generates positive emotions and encourages repetition of the exchange behaviors. Oxytocin is released, which reinforces a sense of bonding and attachment to one's trading partner, thereby enhancing trust and setting off a positive feedback loop of additional trade and trust. Brain scans on subjects playing Prisoner's Dilemma—where cooperation and defection result in differing payoffs depending on what the other participants do—revealed that when subjects were cooperating, the brain areas that lit up were the same regions activated in response to such stimuli as desserts, money, cocaine, and beautiful faces. The neurons most responsive were those rich in dopamine located in the anteroventral striatum in the middle of the brain—the so-called pleasure center. Tellingly, cooperative subjects reported increased feelings of trust toward and camaraderie with like-minded partners.[18] In research with subjects playing a version of the ultimatum game that includes numerous exchanges, when trust is established and the cooperative mode is fully engaged by both trading partners, levels of oxytocin in the blood increase. You can even reverse the causal link—giving subjects a nose spray that includes a dose of oxytocin induces them to cooperate twice as much as they normally would. Trust is good for business and is among the most powerful factors affecting economic growth in a country.[19]

The psychology of trade probably has as much to do with forming alliances between individuals and groups as the economics of trade does in generating an increase in the supply of resources. Nevertheless, the end result of initial exchange activities is an increase in cooperation, mutual aid, and trust, which drives further trade and trust into a positive feedback loop that accentuates amity between people and attenuates enmity between groups, leading to greater peace and prosperity for more people, in more places, more of the time.

✦   ✦   ✦

Bastiat's Principle holds not only for hunter-gatherers but for consumer-traders as well. Note, for example, that in the modern world of consumer-trading nation-states, economic sanctions are among the first steps taken by a nation against another when diplomatic conflict resolution attempts break down. Often such sanctions are imposed for purely economic reasons in a mercantilist mode, as when the United States imposed import tariffs on steel purchased from China and Russia in 2002, which the World Trade Organization declared to be illegal. Economic sanctions are also imposed for political reasons, as when the United States enforced them on Japan after its invasion of China in the 1930s, and these became a prelude (among other factors) to Japan's retaliatory bombing of Pearl Harbor in 1941 and our involvement in the greatest war in history. Or more recently, economic sanctions were imposed by the United States and Japan on India following its 1998 nuclear tests, by the United States on Iran because of the latter's state sponsorship of terrorism, and by the United Nations on Iraq as a tool to force the Iraqi government to comply with U.N. weapons inspectors' search for weapons of mass destruction. Economic sanctions send this message: *if you do not change your behavior, we will no longer trade with you*. And, by Bastiat's Principle, *where our goods do not cross your frontiers, our armies will*. Not inevitably, of course, but often enough in history that the principle retains its veracity. Economic sanctions are not a necessary or even sufficient cause of war, but they are often a prelude to war.

In his books on globalization, the *New York Times* foreign correspondent Thomas Friedman has proposed what he calls the McDonald's Theory of War and the Dell Theory of Conflict Prevention. In the former, Friedman holds, "No two countries that both had McDonald's had fought a war against each other since each got its McDonald's." In the latter, Friedman claims, "No two countries that are both part of a major global supply chain, like Dell's, will ever fight a war against each other as long as they are both part of the same global supply chain."[20] This was more of a literary device to make a point about the power of international trade on human relations than it was a law of social science, since exceptions abound: the 1989 invasion of Panama by the United States, the 1999 bombing of the Federal Republic of Yugoslavia by NATO forces, the constant conflicts between India and Pakistan, and the on-again, off-again clashes between Lebanon and Israel over the past quarter century. All of these countries sell Big Macs. But Friedman's point was that nations that share strong economic ties are less likely to go to war with one another because it raises the stakes even higher than they would otherwise be, and that is my point here in treating this as a probability condition rather than an absolute law.

I made a similar observation on a 2000 trip to Beijing for a scientific conference, during which I toured the ancient Forbidden City complex only to encounter there a brand-new Starbucks café. My *Scientific American* column on the experience, entitled "Starbucks in the Forbidden City," was about the power of science and economics jointly to bring together such disparate Eastern and Western cultures over a scientific discussion or cup of coffee.[21] I call this the Starbucks corollary to Bastiat's Principle: *where Starbucks crosses frontiers, armies will not*. That is, the free trade of products between peoples, and open access to services across geographic borders, obviates the necessity of political borders and thereby decreases the probability that armies will cross them. To the Starbucks corollary I add the Google theory of peace: *where information and knowledge cross frontiers, armies will not*. That is, the free trade of information between peoples, and open access to knowledge across geographic borders, obviates the necessity of political borders and thereby decreases the probability that armies will cross them. A stirring example can be seen in Europe. Since the Treaty of Rome and the formation of the European Union—which integrated disparate and historically divided European nations under one economic umbrella—where once invasions and wars were commonplace throughout a thousand years of European history, they are now unthinkable. Try it. Imagine Germany invading France and waging war upon her, or picture France motoring its armies through the Chunnel and marching them into London to declare the country French. What once made for dramatic literature now sounds like pulp fiction.[22]

The Wikification of the economy—Wikinomics, as it is becoming known[23]—adds to the Google theory of peace the entire world economy as practiced by and participated in by billions of people. Wikipedia is the right analogue for this emerging economic phenomenon. It is the collaboratively created encyclopedia that runs on wiki (Hawaiian for "quick") software that allows real-time and constant editing of documents by anyone, anywhere, anytime. It is an open-sourced, peer-produced, mass-collaborated, bottom-up, self-organized, emergent property of millions of people choosing to build the modern equivalent of the Alexandrian library whose purpose was to make the sum of the world's knowledge available to everyone in one location. Granted, the ancient Alexandrian Greeks had far less knowledge to store than we do today—by many orders of magnitude—but we have the World Wide Web. In the long run, no dictator, demagogue, priest, president, or any other pretender to power will be able to control the Googlification, Wikification, eBayification, MapQuestification, YouTubification, MySpacification of information, knowledge, geography, personal relationships, markets, and the economy.[24] Chinese bureaucrats can attempt to put all the firewalls and controls they want on a billion potential Chinese Web surfers, but they will never be able to prevent knowledge, products, and people from finding their way to those who seek them. *Freedom finds a way.*

✦ ✦ ✦

There is nothing natural about a free economy, and there is nothing inevitable about human groups evolving toward a free market. For thousands of years, chiefdoms, states, and empires practiced slavery and justified it. Ever since the Enlightenment, however, there has been a concerted effort on the part of many states and empires to abolish slavery and promulgate freedom. Several bloody centuries later, slavery has nearly disappeared from the first world, and is rapidly vanishing from the second and third worlds as they transition toward becoming first-world nations.

Despite our natural inclination toward tribalism and xenophobia, we have seen considerable progress with the spread of Enlightenment values, with more rights granted to more people in more places, and with people being protected through both education and legislation from being discriminated against based on such characteristics as race, ethnicity, religion, and gender. Nevertheless, far too many people around the globe still live in political tyranny and economic poverty to the point where they cannot even reach the level of having their primary survival needs met, the sort of needs that allow one to reach a base level of happiness on which to build a meaningful life.

Yet it is not enough simply to oppose slavery, poverty, war, violence, racism, tribalism, and the like. We also have to be *for* something. As Ludwig von Mises warned his fellow anti-Communists at the height of that movement in the 1950s: "An anti-something movement displays a purely negative attitude. It has no chance whatever to succeed. Its passionate diatribes virtually advertise the program they attack. People must fight for something that they want to achieve, not simply reject an evil, however bad it may be."[25] What is it we are fighting for? Freedom. But as we have seen, freedom will not come about on its own, so we must pay a price for it. What is that price?

In the long history of humanity's struggle against the bonds restricting freedom, it would be difficult to find a more eloquent spokesperson than Thomas Jefferson, who understood in the most personal and public sense the toll to open liberty's gate: "The price of freedom is eternal vigilance."[26] It is not inevitable that one day we shall all thrive in an environment of peace, prosperity, and freedom. Given our dual disposition to be both good and evil, and the power of the environment to elicit one or the other, we must *choose* freedom, then create the circumstances in which it can be realized, and then defend it once it is achieved. So freedom begins with an idea, and a conscious choice to attain it. To that end, this entire book is an exercise in consciousness-raising for freedom.

Can the mere raising of people's consciousness work to trigger social change that leads to an increase in freedom? Of course it can. If it couldn't, there would have been no civil rights movement, we'd still be practicing slavery, and women could not vote. How do we get from here to there? Through the slow but steady spread of liberal democracy and free market capitalism, the establishment of environments that spawn interpersonal and international trust, the transparency of political power and economic hegemony, the availability and accessibility of all knowledge for everyone everywhere, and the opening up of political and economic borders, in order—in the words on a plaque posted at the Suez Canal[27]—

Aperire Terram Gentibus

To Open the World to All People

NOTES

Prologue: Economics for Everyone

1. Robert K. Merton, "The Matthew Effect in Science," *Science* 159(38) (1968): 56–63.

2. For example, a relatively unknown sociologist named Marcello Truzzi wrote in an obscure academic journal article about the paranormal that "extraordinary claims require extraordinary evidence." When Carl Sagan repeated the line in his documentary series *Cosmos*, it was thereafter credited to him, and one routinely hears and sees the quote prefaced with "As Carl Sagan said, . . ."

3. As an example of the bestseller effect in publishing, I organize and sponsor a well-attended public science lecture series at Caltech in Pasadena, where we routinely host the biggest names in science, usually when they are in the Los Angeles area on a book tour. We usually order and sell their books for them, but occasionally publishers will request that we arrange to have a bookstore that reports sales to the *New York Times Book Review* sell the books.

4. www.musiclab.columbia.edu.

5. Matthew Salganik, Peter S. Dodds, and Duncan J. Watts, "Experimental Study of Inequality and Unpredictability in an Artificial Cultural Market," *Science* 311 (2006): 854–56. See also Duncan J. Watts, "Is Justin Timberlake a Product of Cumulative Advantage?" *The New York Times Magazine*, April 15, 2007, 22–25.

6. John Tooby and Leda Cosmides, "Friendship and the Banker's Paradox: Other Pathways to the Evolution of Adaptations for Altruism," *Proceedings of the British Academy* 88 (1996): 119–43.

7. Ibid., 134–35.

8. Ibid., 133–34.

9. Letter dated September 18, 1861, in Francis Darwin, *The Life and Letters of Charles Darwin* (London: John Murray, 1887), vol. 2, 121.

10. Michael Shermer, "Colorful Pebbles and Darwin's Dictum," *Scientific American* (April 2001): 38. A parallel to Darwin's Dictum is what I call "Wallace's Wisdom," named after Alfred Russel Wallace, the codiscoverer of natural selection and Darwin's younger contemporary who lived in the penumbra of the Darwinian eclipse. Wallace wrote, "The human mind cannot go on for ever accumulating facts which remain unconnected and without any mutual bearing and bound together by no law." Quoted in John Marchant, *Alfred Russel Wallace, Letters and Reminiscences* (New York: Arno Press, 1916), 63. In my biography of Wallace I employ this principle to emphasize the supreme role of theory in the history of science. Michael Shermer, *In Darwin's Shadow: The Life and Science of Alfred Russel Wallace* (New York: Oxford University Press, 2002), 4.

11. Michael Shermer, *Why Darwin Matters: The Case Against Intelligent Design* (New York: Times Books, 2006).

12. http://www.loc.gov/loc/cfbook/booklists.html. After the Bible and *Atlas Shrugged* were *The Road Less Traveled* by M. Scott Peck, *To Kill a Mockingbird* by Harper Lee, *The Lord of the Rings* by J.R.R. Tolkien, *Gone with the Wind* by Margaret Mitchell, *How to Win Friends and Influence People* by Dale Carnegie, *The Book of Mormon, The Feminine Mystique* by Betty Friedan, *A Gift from the Sea* by Anne Morrow Lindbergh, *Man's Search for Meaning* by Victor Frankl, *Passages* by Gail Sheehy, and *When Bad Things Happen to Good People* by Harold S. Kushner.

13. In my 1997 book, *Why People Believe Weird Things*, I devoted a chapter to the cultlike following that developed around Rand and her philosophy ("The Unlikeliest Cult in History," I called it), in an attempt to show that extremism of any kind, even the sort that eschews cultish behavior, can become irrational. I cited the description of Rand's inner circle by Nathaniel Branden, Rand's chosen intellectual heir, where he listed the central tenets to which followers were to adhere, including "Ayn Rand is the greatest human being who has ever lived. *Atlas Shrugged* is the greatest human achievement in the history of the world. Ayn Rand, by virtue of her philosophical genius, is the supreme arbiter in any issue pertaining to what is rational, moral, or appropriate to man's life on earth. No one can be a good Objectivist who does not admire what Ayn Rand admires and condemn what Ayn Rand condemns. No one can be a fully consistent individualist who disagrees with Ayn Rand on any fundamental issue." Nathaniel Branden, *Judgment Day: My Years with Ayn Rand* (Boston: Houghton Mifflin, 1989), 255–56. Many of the characteristics of a cult, in fact, seemed to fit what the followers of Objectivism believed, most notably veneration of the leader, belief in the infallibility and omniscience of the leader, and commitment to the absolute truth and absolute morality as defined by the belief system.

14. My religious conversion and deconversion are recounted in my book *How We Believe: Science, Skepticism, and the Search for God* (New York: Times Books, 2000).

15. Michael Shermer, "Choice in Rats as a Function of Reinforcer Intensity and Quality: A Thesis Presented to the Faculty of California State University, Fullerton, in Partial Fulfillment of the Requirements for the Degree Master of Arts in Psychology," August 8, 1978.

16. Galambos defined freedom as "the societal condition that exists when every individual has full (i.e., 100%) control over his own property," and a free society as one where "anyone may do anything that he pleases—with no exceptions—so long as his actions affect only his own property; he may do nothing which affects the property of another without obtaining consent of its owner." Galambos never published his long-promised book in his lifetime, so my summary of his theory comes from my own extensive notes

from the V-50 class, and a series of three-by-five leaflets he printed called "Thrust for Freedom," numbered sequentially and presenting the definitions quoted here. In 1999, Galambos's estate issued volume 1 of *Sic Itur Ad Astra* (*The Way to the Stars*), a 942-page tome published by The Universal Scientific Publications Company, Inc. Galambos's dream was to be a space entrepreneur and fly customers to the moon. In his logic, in order to realize this dream he believed that space exploration had to be privatized, which meant that society itself, in its entirety, would have to be privatized. Too bad Galambos did not live long enough to witness the space entrepreneur and libertarian Burt Rutan succeed in being the first to build a private rocket that reached space—it is a lesson libertarians should take to heart.

17. I recount my cycling experiences and the founding of the Ultra-Marathon Cycling Association and the Race Across America in my books *Sport Cycling* (Chicago: Contemporary Books, 1985), and *Race Across America: The Agonies and Glories of the World's Longest and Cruelest Bicycle Race* (Waco, TX: WRS Publishing, 1989).

18. Ludwig von Mises, *Human Action*, 3rd ed. (Chicago: Contemporary Books, 1966), orig. pub. 1949.

19. Baruch Spinoza, *Tractatus Politicus,* edited with an introduction by R. H. M. Elwes, translated by A. H. Gosset (London: G. Bell & Son, 1883), orig. pub. 1667. Emphasis added.

20. Michael Shermer, *Why People Believe Weird Things* (New York: W.H. Freeman, 1997); Shermer, *How We Believe* (New York: Owl Books, 2002); Shermer, *The Science of Good and Evil* (New York: Times Books, 2004).

1. The Great Leap Forward

1. The 100,000-year figure is a low estimate of the range given by paleoanthropologists for when anatomically modern humans migrated out of Africa and began to colonize Europe and the rest of the world, usually given as a range between 100,000 and 160,000 years ago. See Timothy D. White, B. Asfaw, D. Degusta, H. Gilbert, G. D. Richards, G. Suwas, and F. Clark Howell, "Pleistocene *Homo sapiens* from Middle Awash, Ethiopia," *Nature* 423 (2003): 742–47. Many hominid species lived in small bands for many millions of years in Africa, but for our purposes the 100,000-year figure will suffice to make the point. For an overview see Richard G. Klein, *The Human Career: Human Biological and Cultural Origins* (Chicago: University of Chicago Press, 1999).

2. Through my friendship with Napoleon Chagnon, the anthropologist whose ethnology *Yanomamö: The Fierce People* (New York: Harcourt Brace, 1992) put them on the world map, I have long been interested in the Yanomamö and have written about their religion in *How We Believe* (New York: Owl Books, 2002), their moral and ethical systems in *The Science of Good and Evil* (New York: Times Books, 2004), and the controversy over Chagnon's ethnography and whether or not they are really the "fierce people" in *Science Friction* (New York: Times Books, 2005). The direct comparison with New Yorkers was made by Eric Beinhocker in *The Origin of Wealth: Evolution, Complexity, and the Radical Remaking of Economics* (Cambridge, MA: Harvard Business School Press, 2006). Beinhocker computes a figure of $93 annual income per person for hunter-gatherers from data collected on GDP by Bradford DeLong, available at http://www.j-bradford-delong.net/. Since the figure is in 1990 dollars, and is an estimate in any case, I rounded up to an even $100 for easy comparison. The estimates for New Yorkers' average income is from the state's government statistics—Beinhocker quotes a mean of $36,000

and a median of $43,160, so I split the difference and rounded off to an even $40,000, again for simple comparison. Beinhocker estimates the Yanomamö SKU figure from Chagnon's ethnography, and the Manhattan SKU figure from the universal product code (UPC) system whose 10-digit (one billion) accounting is now full and the system is switching to a 13-digit (one trillion) code. I'm not sure why he uses a 10-billion figure instead of one billion, but whether the SKU difference between hunter-gatherers and consumer-traders is seven or eight orders of magnitude does not really matter—the point is made regardless of the precise accuracy of the figures.

3. Once again, these are back-of-the-envelope calculations for comparison purposes only. The average male walks at about 3.5 miles per hour, or 5.67 kilometers per hour, so it would take 261 hours to walk 1,480 kilometers, which is just under eleven days, assuming you didn't stop to eat, rest, or take care of other necessities. The distance between Earth and Jupiter varies considerably depending on where in our respective orbits we are when the measurement is taken. The speed of the Voyager I spacecraft has also varied, most notably in its acceleration through the "slingshot" effect of receiving a "gravity boost" from planets it approached. My figure of about 51,000 kilometers per hour comes from the fact that it took Voyager I a year and a half to reach Jupiter (it was launched on September 5, 1977, and arrived at Jupiter on March 5, 1979). It is now traveling at about 63,000 kilometers per hour, and on August 12, 2006, it reached the heliosheath, the termination shock region between our solar system and interstellar space; that is, the zone where our sun's influence gives way to interstellar space and the influence of interstellar gas and other stars. Voyager I is now traveling at about 538,552,332 kilometers per year and is now about 15 billion kilometers from Earth. Even at this almost incomprehensible speed, it would take 74,912 years to get to the Alpha Centauri star system, the closest stars to our sun, if it were heading in that direction, which it is not.

4. David Chalmers, *The Conscious Mind: In Search of a Fundamental Theory* (New York: Oxford University Press, 1997).

5. Marvin Minsky, *Society of Mind* (New York: Simon and Schuster, 1988). We know that the mind is a product of the brain, but how? We know that conscious thought is produced by neurons firing, but how? No one fully understands what is called the Neural Correlates of Consciousness, but much progress has been made in recent years by modeling neural networks, or the actions of networks of neurons out of which emerges a more complex phenomenon we call mind or consciousness. See Christof Koch, *The Quest for Consciousness: A Neurobiological Approach* (Denver, CO.: Roberts & Co., 2004).

6. Personal correspondence between myself, Dawkins, and Hardison. See also vol. 9, no. 4 of *Skeptic*, that presents the details of these computer experiments and what they mean for how evolution works, as well as the original works in which they were published: Richard Hardison, *Upon the Shoulders of Giants* (Baltimore: University Press of America, 1985); and Richard Dawkins, *The Blind Watchmaker* (New York: W. W. Norton, 1986). Hardison's response to Dawkins's revelation was equally insightful on the power of this model:

Incidentally, I never felt that the TOBEORNOTTOBE example was entirely original with me. Bob Newhart, the comic, did a very nice skit in which he proposed an infinite number of monkeys working with an infinite number of typewriters, and then he realized that he would also need an infinite number of "inspectors" looking over the shoulders of the monkeys to see if anything meaningful occurred. Newhart then put himself into the role of one of these inspectors, spending another boring day and finding nothing. "Dum de dum de dum . . . Boring . . . Oh . . . Hey, Charlie, I think I have one. Let's see, yeah. 'To Be Or Not To Be, that is the acxrotphoeic.'" I simply realized that Bob's humor might be a useful way of helping students to comprehend the selective nature of the "struggle for survival." So you see that my contribution was minimal.

7. I am grateful to David B. Schlosser for some of these examples as well as his insights into the evolutionary basis of economics from his very real-world experiences as a businessman and a congressional candidate.

8. Ludwig von Mises, *Socialism* (Indianapolis: Liberty Classics, 1981). See also Murray Rothbard, *The Essential Ludwig von Mises* (Auburn, AL: The Ludwig von Mises Institute of Auburn University, 1980).

9. R. Preston McAfee, "The Price Is Right Mysterious," *Engineering and Science* 3 (2005): 32–42; Michael Doane, Kenneth Hendricks, and R. Preston McAfee, "Evolution of the Market for Air-Travel Information," http://vita.mcafee.cc/PDF/AirTravel.pdf, 2003; Joseph Turow, Lauren Feldman, and Kimberly Meltzer, "Open to Exploitation: American Shoppers Online and Offline," policy statement, Annenberg Policy Center, University of Pennsylvania, 2005.

10. Ultimatum game research and applications are reviewed in Colin Camerer, *Behavioral Game Theory* (Princeton, NJ: Princeton University Press, 2003).

11. Herbert Gintis, Samuel Bowles, Robert Boyd, and Ernst Fehr, *Moral Sentiments and Material Interests: The Foundations of Cooperation in Economic Life* (Cambridge, MA: MIT Press, 2005).

12. Frans B. M. de Waal, "Food-Transfers Through Mesh in Brown Capuchins," *Journal of Comparative Psychology,* 111 (1997): 370–78; and "Food Sharing and Reciprocal Obligations Among Chimpanzees," *Journal of Human Evolution* (1989): 433–59.

13. James Madison, "The Federalist No. 51: The Structure of the Government Must Furnish the Proper Checks and Balances Between the Different Departments," *Independent Journal* (Wednesday, February 6, 1788).

2. Our Folk Economics

1. Richard Dawkins, *The God Delusion* (New York: Houghton Mifflin, 2006), 367–68.

2. K. Hawkes, "Showing Off: Tests of an Hypothesis about Men's Foraging Goals," *Ethnology and Evolutionary Biology* 12 (1990): 29–54; Hillard Kaplan and Kim Hill, "Food Sharing Among Ache Foragers: Tests of Explanatory Hypotheses," *Current Anthropology* 26 (1985): 223–46.

3. P. Freuchen, *Book of the Eskimos* (Cleveland: World Publishing, 1961).

4. For a brilliant and highly readable history of the transition from the zero-sum interactions of our ancestors to the nonzero world of today, see Robert Wright, *Nonzero: The Logic of Human Destiny* (New York: Pantheon, 2000).

5. Quoted in Mark Skousen, *The Making of Modern Economics* (London: M. E. Sharpe, 2001), 20.

6. Daniel B. Klein, "The People's Romance: Why People Love Government (As Much As They Do)," *The Independent Review* X(1) (Summer 2005): 5–37.

7. Personal correspondence, March 9, 2007.

8. I provide numerous examples of this bias in the next chapter. Please note that I am using the terms "conservative" and "liberal" in their modern usages, and recognize

that the eighteenth-century term "classical liberal" refers to those who tend to favor free markets.

9. Stephen Jay Gould, "Kropotkin Was No Crackpot," *Natural History* (July 1998): 12–21.

10. Peter A. Corning, "Evolutionary Ethics: An Idea Whose Time Has Come? An Overview and an Affirmation," *Politics and the Life Sciences* 22(1) (2003): 50–77.

11. Pyotr Kropotkin, *Mutual Aid: A Factor in Evolution* (London: Heinemann, 1902).

12. Daniel P. Todes, "Darwin's Malthusian Metaphor and Russian Evolutionary Thought, 1859–1917," *Isis* 78(294) (1987): 537–51.

3. Bottom-Up Capitalism

1. Janet Browne, *Voyaging: Charles Darwin—A Biography* (New York: Knopf, 2000), 36, 366.

2. Toni Vogel Carey, "The Invisible Hand of Natural Selection, and Vice Versa," *Biology & Philosophy* 13(3) (July 1998): 427–42; Michael T. Ghiselin, *The Economy of Nature and the Evolution of Sex* (Berkeley: University of California Press, 1974); Stephen Jay Gould, "Darwin's Middle Road," in *The Panda's Thumb* (New York: W.W. Norton, 1980), 59–68; Stephen Jay Gould, "Darwin and Paley Meet the Invisible Hand," in *Eight Little Piggies* (New York: W. W. Norton, 1993), 138–52; Elias L. Khalil, "Evolutionary Biology and Evolutionary Economics," *Journal of Interdisciplinary Economics* 8(4) (1997): 221–44; Silvan S. Schweber, "Darwin and the Political Economists: Divergence of Character," *Journal of the History of Biology* 13 (1980): 195–289; Syed Ahmad, "Adam Smith's Four Invisible Hands," *History of Political Economy* 22(1) (Spring 1990): 137–44; Donald Walsh, "Darwin Fallen Among Political Economists," *Proceedings of the American Philosophical Society* 145(4) (2001): 415–37.

3. William Paley, *Natural Theology: Or, Evidences of the Existence and Attributes of the Deity, Collected from the Appearances of Nature* (Hallowell, ME: Glazier & Co., 1826), 169–71 (orig. pub. 1802). Emphasis added.

4. Adam Smith, *The Theory of Moral Sentiments* (London: A. Millar, 1759), part 1, sec. 1, chap. 1.

5. Ibid., part 1, sec. 1, chap. 40.

6. Adam Smith, *An Inquiry into the Nature and Causes of the Wealth of Nations*, 2 vols., R. H. Campbell and A. S. Skinner, gen. eds., W. B. Todd, text ed. (Oxford: Clarendon Press, 1979), 549 (orig. pub. 1776).

7. Ibid., 131.

8. Ibid., 625.

9. Ibid., 424.

10. Ibid., 625.

11. http://www.taxfoundation.org/.

12. Frédéric Bastiat, "The Petition of the Candlemakers" and "What Is Seen and What Is Not Seen," in *Selected Essays on Political Economy*, George B. de Huszar, ed. (Irvington-on-Hudson, NY: Foundation for Economic Education, 1995) (orig. pub. 1845 and 1848).

13. Bastiat, "What Is Seen and What Is Not Seen," chap. 1.

14. Frédéric Bastiat, *Economic Sophisms* (Irvington-on-Hudson, NY: Foundation for Economic Education, 1996) (orig. pub. 1845).

15. Richard D. Stone, *The Interstate Commerce Commission and the Railroad Industry: A History of Regulatory Policy* (New York: Praeger, 1991).

16. The complete text of the Sherman Antitrust Act is available at the following government Web page: http://www.usdoj.gov/atr/foia/divisionmanual/ch2.htm#a1. See also Dominick Armentano, *Antitrust and Monopoly: Anatomy of a Policy Failure* (New York: Wiley, 1982); and Yale Brozen, *Concentration, Mergers, and Public Policy* (New York: Macmillan, 1982).

17. I am grateful to Jay Stuart Snelson and his seminar on free market economics taught through his Institute for Human Progress for making the distinction between consumer and producer driven economics in Adam Smith's theory, and for the example of Alcoa as a violation of consumer driven economics, to which I added my own spin of distinguishing between bottom-up and top-down economics.

18. U.S. v. Aluminum Co. of America, 148 F.2d 416, 431 (2d Cir. 1945).

19. The former chairman of the Federal Reserve Board, Alan Greenspan, penned a succinct history and summation of the effects of antitrust legislation in his 1966 essay "Antitrust," in *Capitalism: The Unknown Ideal*, Ayn Rand, ed. (New York: Signet, 1966). There is no question that government subsidies of the railroads opened them up to subsequent strings tied to those subsidies in the form of legislation regulating that (and subsequent) industries.

20. U.S. v. Microsoft, No. 98–1232 (TPJ) (D.D.C. November 5, 1999) (Court's Findings of Fact), paragraph 408. The entire document is available on the U.S. government Web page http://www.usdoj.gov/atr/cases/f3800/msjudgex.htm.

21. Ibid., paragraph 409.

22. Ibid., paragraph 375.

23. AP wire story, "Motorcycle Imports Cited," *New York Times*, January 20, 1983.

24. United States International Trade Commission, Heavyweight Motorcycles, and Power Train Subassemblies Therefor: Report to the President on Investigation No. TA-201-4F under Section 201 of the Trade Act of 1974, February 1983, 19. After the first year rate of 49.4 percent, the tariff was lowered to 39.4 percent in the second year, 24.4 percent in the third year, 19.4 percent in the fourth year, and 14.4 percent in the fifth year. After that the tariff would return to its original 4.4 percent.

25. George Will, "Liberalism as Condescension," *Washington Post*, September 14, 2006. Available at http://www.realclearpolitics.com/articles/2006/09/liberalism_as_condescension.html.

26. Edward C. Prescott, "Competitive Cooperation," *Wall Street Journal*, Opinion, Feb. 15, 2007, A19.

27. Ibid.

28. Mary Anastasia O'Grady, "One Year After CAFTA," *Wall Street Journal*, February 26, 2007, A18.

29. Prescott, "Competitive Cooperation."

30. Smith, *Wealth of Nations*, 145.

31. Ibid., 418.

32. Ibid., 14.

33. Ibid., 423. Emphasis added. For a lengthy discussion of the origins of the invisible hand metaphor, see Emma Rothschild, *Economic Sentiments* (Cambridge: Harvard University Press, 2001). Emma Rothschild notes that Smith used the invisible hand metaphor in an early work on the history of astronomy, in a discussion of how polytheistic societies lead people to attribute "the irregular events of nature" to "intelligent, though invisible beings—to gods, demons, witches, genii, fairies." They do not ascribe divine support to "the ordinary course of things": "Fire burns, and water refreshes; heavy bodies descend, and lighter substances fly upwards, by the necessity of their own

nature; nor was the invisible hand of Jupiter ever apprehended to be employed in those matters." Rothschild also suggests that Smith may originally have picked up the metaphor from Shakespeare, in *Macbeth: Come, seeling night, / Scarf up the tender eye of pitiful day, / And with thy bloody and invisible hand / Cancel and tear to pieces that great bond / Which keeps me pale.* There is, however, no direct proof of a connection between Smith and Shakespeare.

34. Charles Darwin, *The Origin of Species by Means of Natural Selection: or The Preservation of Favoured Races in the Struggle for Life* (London: Charles Murray, 1859), 133.

35. Darwin most likely made the connection between the two economies early in the development of his theory—around 1838—when he read Dugald Stewart's *On the Life and Writing of Adam Smith*, in such passages as this: "The most effective plan for advancing a people . . . is by allowing every man, as long as he observes the rules of justice, to pursue his own interest in his own way, and to bring both his industry and his capital into the freest competition with those of his fellow citizens. Every system of policy which endeavors . . . to draw towards a particular species of industry a greater share of the capital of the society than would naturally go to it . . . is, in reality, subversive of the great purpose which it means to promote." Read "animal" for "man" and "population" for "people," and we have a perfect description of natural selection in nature. See Silvan S. Schweber, "The Origin of the *Origin* Revisited," *Journal of the History of Biology* 10 (1977): 229–316. Schweber argues that Darwin put the pieces of his theory together only after extensive reading of works outside of natural history, such as the Scottish economist Adam Smith: "The Scottish analysis of society contends that the combined effect of individual actions results in the institutions upon which society is based, and that such a society is a stable and evolving one and functions without a designing and directing mind."

36. Listen to Darwin's description of the evolution of the social instincts and our natural inclination to feel sympathy for others and how this led to the development of a moral system, from his 1871 *The Descent of Man*, and you will hear the echo of Adam Smith's *A Theory of Moral Sentiments*: "The following proposition seems to me in a high degree probable—namely, that any animal whatever, endowed with well marked social instincts, the parental and filial affections being here included, would inevitably acquire a moral sense or conscience, as soon as its intellectual powers had become as well, or nearly as well developed, as in man." And: "The social instincts, which no doubt were acquired by man as by the lower animals for the good of the community, will from the first have given to him some wish to aid his fellows, some feeling of sympathy, and have compelled him to regard their approbation and disapprobation. Such impulses will have served him at a very early period as a rude rule of right and wrong." Charles Darwin, *The Descent of Man, and Selection in Relation to Sex*, 2 vols. (London: John Murray, 1871), vol. 1, 71–72.

4. Of Pandas, Products, and People

1. Stephen Jay Gould, *Wonderful Life: The Burgess Shale and the Nature of History* (New York: W. W. Norton, 1989), 283.

2. Edward Lorenz, "Predictability: Does the Flap of a Butterfly's Wings in Brazil Set Off a Tornado in Texas?" Address at the AAAS annual meeting, Washington, D.C., December 29, 1979.

3. Stephen Jay Gould, "The Panda's Peculiar Thumb," *Natural History* 9 (1978): 20–30.

4. Paul A. David, "Clio and the Economics of QWERTY," *American Economic Review* 75 (1985): 332–37; see also his "Understanding the Economics of QWERTY: The

Necessity of History," in *Economic History and the Modern Economist*, W. N. Parker, ed. (London: Blackwell, 1985); and "Path Dependence in Economic Processes: Implications for Policy Analysis in Dynamical System Contexts," in *The Evolutionary Foundations of Economics*, Kurt Dopfer, ed. (New York: Cambridge University Press, 2005), 151–94.

5. David, "Clio and the Economics of QWERTY."

6. John Nash, "Equilibrium Points in N-Person Games," *Proceedings of the National Academy of Sciences* 36(1) (1950): 48–49; and "Non-Cooperative Games," *The Annals of Mathematics* 54(2) (1951): 286–95. See also Oskar Morgenstern and John von Neumann, *The Theory of Games and Economic Behavior* (Princeton, NJ: Princeton University Press, 1947).

7. For a general discussion of the most recent research on game theory in light of such concepts as Nash equilibrium and Pareto optimization, see Colin F. Camerer, *Behavioral Game Theory: Experiments in Strategic Interaction* (Princeton, NJ: Princeton University Press, 2003).

8. John Maynard Smith, *Evolution and the Theory of Games.* (Cambridge: Cambridge University Press, 1982); and Maynard Smith and Eros Szathmary, The *Major Transitions in Evolution* (Oxford University Press, 1998).

9. For the history of the QWERTY keyboard, see Paul David, "Understanding the Economics of QWERTY: The Necessity of History"; and Stephen Jay Gould, "The Panda's Thumb of Technology," *Natural History* (January 1984): 14–23. For a general history of the typewriter see F. T. Masi, ed., *The Typewriter Legend* (Secaucus, NJ: Matsushita Electric Corp. of America, 1985); F. J. Romano, *Machine Writing and Typesetting* (Salem, MA: GAM Communications, 1986); and D. R. Hoke, *Ingenious Yankees: The Rise of the American System of Manufactures in the Private Sector* (New York: Columbia University Press, 1990). Hoke notes the paucity of historical records for reconstructing the history of the typewriter and was forced to rely on company histories, advertisements from magazines, photographs and illustrations of typewriters, surviving typewriters, and biographical material on the inventors, manufacturers, and entrepreneurs in the industry.

10. Stanley J. Liebowitz and Stephen E. Margolis, "The Fable of the Keys," *Journal of Law and Economics* 33 (April 1990); Liebowitz and Margolis, *The Economics of Qwerty*, papers by Stanley Liebowitz and Stephen Margolis; Peter Lewin, ed. (MacMillan/NYU Press, 2002); Margolis with Liebowitz, "Path Dependence, Lock-in and History," *Journal of Law, Economics and Organization* (Summer 1995): 205–26; and Liebowitz and Margolis, "Typing Errors," *Reason* (June 1996).

11. Liebowitz and Margolis, "Typing Errors."

12. Ibid.

13. W. Brian Arthur, "Positive Feedbacks in the Economy," *Scientific American* 262 (1990): 92–99.

14. David, "Clio and the Economics of QWERTY" (see note 4 above).

15. Paul Krugman, "QWERTY, Lock-In, and Path Dependence," 2001. Accessed at http://cscs.umich.edu/~crshalizi/notebooks/qwerty.html.

16. Douglas J. Puffert, "Path Dependence in Spatial Networks: The Standardization of Railway Track Gauge," *Explorations in Economic History* 39(3) (July 2002): 282–314.

17. George Basalla, *The Evolution of Technology* (Cambridge: Cambridge University Press, 1988).

18. Quoted in Basalla, *Evolution*, 53.

19. Ibid., 30.

20. Ibid., 123.

21. Ibid., 128.

22. Charles Darwin, *On the Various Contrivances by Which British and Foreign Orchids Are Fertilized by Insects, and on the Good Effects of Intercrossing* (London: John Murray, 1862), 348.

23. Stephen Jay Gould and Elizabeth Vrba, "Exaptation: A Missing Term in the Science of Form," *Paleobiology* 8 (1982): 4–15.

24. R. O. Prum and A. H. Brush, "Which Came First, the Feather or the Bird: A Long-Cherished View of How and Why Feathers Evolved Has Now Been Overturned," *Scientific American* (March 2003): 84–93.

25. Kevin Padian and L. M. Chiappe, "The Origin of Birds and Their Flight," *Scientific American* (February 1998): 38–47. In the Galápagos islands, for example, I have photographed flightless cormorants returning to shore after diving for food in the sea, upon which they stretch out their stubby little wings with desultory feathers to dry them out and collect heat from the sun. In this case, the exaptation was from flight to thermoregulation. The Galápagos are also home to a species of penguin whose wings have been exapted for propulsion and steering in the water.

26. K. P. Dial, "Wing-Assisted Incline Running and the Evolution of Flight," *Science* 299 (2003): 402–4; P. Burgers and L. M. Chiappe, "The Wing of Archaeopteryx as a Primary Thrust Generator," *Nature* 399 (1999): 60–62; P. Burgers and Kevin Padian, "Why Thrust and Ground Effect Are More Important Than Lift in the Evolution of Sustained Flight," in *New Perspectives on the Origin and Evolution of Birds: Proceedings of the International Symposium in Honor of John H. Ostrom*, J. Gauthier and L. F. Gall, eds. (New Haven, CT: Peabody Museum of Natural History, 2001), 351–61.

27. Alan Gishlick, "Evolutionary Paths to Irreducible Systems: The Avian Flight Apparatus," in *Why Intelligent Design Fails*, Matt Young and Taner Edis, eds. (New Brunswick: Rutgers University Press, 2004), 58–71.

28. Stephen Jay Gould, "Tires to Sandals," *Natural History* (April 1989): 8–15.

29. As a final example of exaptation on a grand scale, bicycle technology has been utilized in, and even spawned, new technologies and industries. Consider this short list compiled by David Gordon Wilson: mass production and use of ball bearings, production and use of steel tubes, use of metal stamping in production, differential gearing, tangent-spoked wheels (later used in cars, motorcycles, and airplanes), bushed power transmission chain, mass production and use of pneumatic tires, good-roads movement, Harley and Davidson, bicycle racers, Wright brothers, bicycle manufacturers, the underpinnings of the automobile age. David Gordon Wilson, *Bicycling Science*, 3rd ed. (Cambridge, MA: MIT Press, 2004), 32.

Trevor Pinch and W. E. Bijker make a similar point: "It may be useful to state explicitly that we consider bicycles to be as fully fledged a technology as, for example, automobiles or aircraft. It may be helpful for readers from outside notorious cycle countries such as the Netherlands, France, and Great Britain to point out that both the automobile and the airplane industries are, in a way, descendants from the bicycle industry. Many names occur in the histories of both the bicycle and the autocar: Triumph, Rover, Humber, and Raleigh, to mention but a few. The Wright brothers both sold and manufactured bicycles before they started to build their flying machines—mostly made out of bicycle parts." T. J. Pinch and W. E. Bijker, "The Social Construction of Facts and Artifacts: Or How the Sociology of Science and the Sociology of Technology Might Benefit Each Other," in *The Social Construction of Technological Systems*, W. E. Bijker, T. P. Hughes, and T. P. Pinch, eds. (Cambridge, MA: MIT Press, 1989), 50.

5. Minding Our Money

1. Leon Festinger, Henry W. Riecken, and Stanley Schachter, *When Prophecy Fails: A Social and Psychological Study* (New York: HarperCollins, 1964), 3.

2. Ibid., 194.

3. By January 9, 1955, Mrs. Keech's group had completely disbanded, and, fittingly, she moved from the Chicago area to Arizona, where she joined another UFO cult—Scientology.

4. For a popular treatment of the subject see Carl Sagan, *The Demon-Haunted World: Science as a Candle in the Dark* (New York: Random House, 1996). For a scholarly treatment of the subject, see Diana Tumminia, *When Prophecy Never Fails: Myth and Reality in a Flying-Saucer Group* (New York: Oxford University Press, 2006).

5. Dan Eggen and Paul Kane, "Gonzales: 'Mistakes Were Made,'" *Washington Post*, March 14, 2007, A01.

6. Carol Tavris and Elliot Aronson, *Mistakes Were Made (but not by me)* (New York: Harcourt, 2007). The quotes from Kissinger, Egan, and the McDonald's spokesperson are cited on page 1.

7. Philip Tetlock, *Expert Political Judgment: How Good Is It? How Can We Know?* (Princeton, NJ: Princeton University Press, 2005).

8. Geoffrey Cohen, "Party over Policy: The Dominating Impact of Group Influence on Political Beliefs," *Journal of Personality and Social Psychology* 85 (2003): 808–82.

9. John Jost and Orsolya Hunyady, "Antecedents and Consequences of System-Justifying Ideologies," *Current Directions in Psychological Science* 14 (2005): 260–65; Aaron C. Kay and John T. Jost, "Complementary Justice: Effects of 'Poor But Happy' and 'Poor But Honest' Stereotype Exemplars on System Justification and Implicit Activation of the Justice Motive," *Journal of Personality and Social Psychology* 85 (2003): 823–37; see also Stephanie Wildman, ed., *Privilege Revealed: How Invisible Preference Undermines America* (New York: New York University Press, 1996).

10. Tavris and Aronson, 130–32. See also http://www.innocenceproject.org/.

11. Paul Ekman, *Telling Lies: Clues to Deceit in the Marketplace, Marriage, and Politics* (New York: W. W. Norton, 1992); and *Emotions Revealed: Recognizing Faces and Feelings to Improve Communication and Emotional Life* (New York: Times Books, 2003).

12. Daniel J. Simons and Christopher Chabris, "Gorillas in Our Midst: Sustained Inattentional Blindness for Dynamic Events," *Perception* 28 (1999): 1059–74. You can watch the video clip at http://viscog.beckman.uiuc.edu/djs_lab/.

13. Simons told me that he has had the same experience: "We actually rewound the videotape to make sure subjects knew we were showing them the same clip." Personal interview, January 8, 2004.

14. Personal interview, January 8, 2004. Simons added, "The mistaken belief that important events will automatically draw attention is exactly why these findings are surprising; it is also what gives them some practical implications. By taking for granted that unexpected events will be seen, people often are not as vigilant as they could be in actively anticipating such events."

15. Emily Pronin, D. Y. Lin, and L. Ross, "The Bias Blind Spot: Perceptions of Bias in Self Versus Others," *Personality and Social Psychology Bulletin* 28 (2002): 369–81.

16. Emily Pronin, Thomas Gilovich, and L. Ross, "Objectivity in the Eye of the Beholder: Divergent Perceptions of Bias in Self Versus Others," *Psychological Review* 111 (2004): 781–99.

17. Personal correspondence, January 7, 2004.

18. S. N. Brenner and E. A. Molander, "Is the Ethics of Business Changing?" *Harvard Business Review* (January/February 1977): 57–71.

19. P. A. M. Van Lange, T. W. Taris, and R. Vonk, "Dilemmas of Academic Practice: Perceptions of Superiority Among Social Psychologists," *European Journal of Social Psychology* 27 (1997): 675–85.

20. J. Kruger, "Personal Beliefs and Cultural Stereotypes About Racial Characteristics," *Journal of Personality and Social Psychology* 71 (1996): 536–48.

21. "Oprah: A Heavenly Body? Survey Finds Talk-Show Host a Celestial Shoo-in," *U.S. News & World Report*, March 31, 1997, 18.

22. M. Ross and F. Sicoly, "Egocentric Biases in Availability and Attribution," *Journal of Personality and Social Psychology* 37 (1979): 322–36; R. M. Arkin, H. Cooper, and T. Kolditz, "A Statistical Review of the Literature Concerning the Self-serving Bias in Interpersonal Influence Situations," *Journal of Personality* 48 (1980): 435-48; and M. H. Davis and W. G. Stephan, "Attributions for Exam Performance," *Journal of Applied Social Psychology* 10 (1980): 235–48. For a general summary of the attribution bias, see Carol Tavris and Carole Wade, *Psychology in Perspective*, 2nd ed. (New York: Longman / Addison Wesley, 1997).

23. R. E. Nisbett and L. Ross, *Human Inference: Strategies and Shortcomings of Social Judgment* (Englewood Cliffs, NJ: Prentice-Hall, 1980).

24. Preliminary results of our study were originally published in my book *How We Believe: The Search for God in an Age of Science* (New York: W. H. Freeman, 2000).

25. The full data set and analysis will be published in Michael Shermer and Frank J. Sulloway, "Religion and Belief in God: An Empirical Study." In preparation.

26. Daniel Kahneman, "Autobiography," Nobel Prize.org: http://nobelprize.org/nobel_prizes/economics/laureates/2002/kahneman-autobio.html (2002).

27. Thomas Gilovich, Richard Vallone, and Amos Tversky, "The Hot Hand in Basketball: On the Misperception of Random Sequences," *Cognitive Psychology* 17 (1985): 295–314.

28. Quoted in obit released by Stanford University and available at http://news-service.stanford.edu/pr/96/960605tversky.html.

29. For a literature review on this and dozens of other problems studied in this area, see Daniel Kahneman, Paul Slovic, and Amos Tversky, eds., *Judgment Under Uncertainty: Heuristics and Biases* (New York: Cambridge University Press, 1982); and, more recently, Thomas Gilovich, Dale Griffin, and Daniel Kahneman, *Heuristics and Biases: The Psychology of Intuitive Judgment* (New York: Cambridge University Press, 2002).

30. Daniel Kahneman and Amos Tversky, "On the Psychology of Prediction," *Psychological Review* 80 (1973): 237–51.

31. Amos Tversky and Daniel Kahneman, "Availability: A Heuristic for Judging Frequency and Probability," in Kahneman, Slovic, and Tversky, *Judgment Under Uncertainty*, 163.

32. Amos Tversky and Daniel Kahneman, "Extension Versus Intuitive Reasoning: The Conjunction Fallacy in Probability Judgment," *Psychological Review* 90 (1983): 293–315.

33. J. S. Carroll, "The Effect of Imagining an Event on Expectations for the Event: An Interpretation in Terms of the Availability Heuristic," *Journal of Experimental Social Psychology* 14 (1978): 88–96.

34. Amos Tversky and Daniel Kahneman, "Availability: A Heuristic for Judging Frequency and Probability," *Cognitive Psychology* 5 (1973): 207–32.

35. B. Combs and P. Slovic, "Newspaper Coverage of Causes of Death," *Journalism Quarterly* 56 (1979): 837–43.

36. Barry Glassner, *The Culture of Fear: Why Americans Are Afraid of the Wrong Things* (New York: Basic Books, 1999).

37. Baruch Fischhoff, "For Those Condemned to Study the Past: Heuristics and Biases in Hindsight," in Kahneman, Slovic, and Tversky, *Judgment Under Uncertainty*, 335–351.

38. John C. Zimmerman, "Pearl Harbor Revisionism," *Intelligence and National Security* 17(2) (2002): 127–46.

39. Colin F. Camerer, George Loewenstein, and Matthew Rabin, eds., *Advances in Behavioral Economics* (Princeton, NJ: Princeton University Press, 2001), 11.

40. Marilyn vos Savant was bombarded with angry letters when she revealed the correct solution in her column: "Ask Marilyn," *Parade*, Sept. 9, 1990; Feb. 17, 1991; July 7, 1991. You can actually play the three-door game at http://utstat.toronto.edu/david/MH .html#1. And on other Web sites you can find computer programs that have run hundreds of thousands of simulations of the game, proving that in the long run it is better to switch doors. See also L. Gillman, "The Car and the Goats," *American Mathematical Monthly* (January 1992): 3–7.

41. Amos Tversky and Daniel Kahneman, "The Framing of Decisions and the Psychology of Choice," *Science* 211 (1981): 453–58; and "Rational Choice and the Framing of Decisions," *Journal of Business* 59(4) (1986): 2; Benededetto De Martino et al., "Frames, Biases, and Rational Decision-Making in the Human Brain," *Science* 313 (2006): 684–87.

42. Drazen Prelec and Duncan Simester, "Always Leave Home Without It: A Further Investigation of the Credit-Card Effect on Willingness to Pay," *Marketing Letters* 12(1) (2001): 5–12.

43. W. Samuelson and R. J. Zeckhauser, "Status Quo Bias in Decision Making," *Journal of Risk and Uncertainty* 1 (1988): 7–59; Daniel Kahneman, J. L. Knetsch, and Richard H. Thaler, "Anomalies: The Endowment Effect, Loss Aversion, and Status Quo Bias," *Journal of Economic Perspectives* 5(1) (1991): 193–206; and E. J. Johnson, J. Hershey, J. Meszaros, and H. Kunreuther, "Framing, Probability Distortions, and Insurance Decisions," *Journal of Risk and Uncertainty* 7 (1993): 35–51.

44. Samuelson and Zeckhauser, "Status Quo Bias in Decision Making."

45. Richard Thaler, Daniel Kahneman, and Jack Knetsch, "Experimental Tests of the Endowment Effect and the Coase Theorem," *Journal of Political Economy* (December 1990).

46. Costs, deaths, and casualties of the Iraq war: http://www.cnn.com/SPECIALS/ 2003/iraq/forces/casualties/. Bush quote: http://www.whitehouse.gov/news/releases/2006/ 07/20060704.html. Clinton quote: http://economistsview.typepad.com/economistsview/ 2006/09/not_quite_ready.html.

47. Raymond Nickerson, "Confirmation Bias: A Ubiquitous Phenomenon in Many Guises," *Review of General Psychology* 2(2) (1998): 175–220.

48. Mark Snyder, "Seek and Ye Shall Find: Testing Hypotheses About Other People," in E. T. Higgins, C. P. Heiman, and M. P. Zanna, eds., *Social Cognition: The Ontario Symposium on Personality and Social Psychology* (Hillsdale, NJ: Erlbaum, 1981), 277–303.

49. John M. Darley and Paul H. Gross, "A Hypothesis-Confirming Bias in Labelling Effects," *Journal of Personality and Social Psychology* 44 (1983): 20–33.

50. Bonnie Sherman and Ziva Kunda, "Motivated Evaluation of Scientific Evidence," paper presented at the annual meeting of the American Psychological Society, Arlington, VA, 1989.

51. Deanna Kuhn, "Children and Adults as Intuitive Scientists," *Psychological Review* 96 (1989): 674–89.

52. Deanna Kuhn, M. Weinstock, and R. Flaton, "How Well Do Jurors Reason? Competence Dimensions of Individual Variation in a Juror Reasoning Task," *Psychological Science* 5 (1994): 289–96.

53. D. Westen, C. Kilts, P. Blagov, K. Harenski, and S. Hamann, "The Neural Basis of Motivated Reasoning: An fMRI Study of Emotional Constraints on Political Judgment During the U.S. Presidential Election of 2004," *Journal of Cognitive Neuroscience* 18 (2006): 1947–58.

54. Thomas Gilovich and Gary Belsky, *Why Smart People Make Big Money Mistakes and How to Correct Them: Lessons from the New Science of Behavioral Economics* (New York: Fireside, 2000).

55. My preferred strategy, which I have yet to find a casino to accept, would be to give the dealer $500 and ask if I can just play for an hour—they're going to get my money anyway, and I am primarily just a recreational gambler.

6. The Extinction of *Homo Economicus*

1. Michael Shermer, "Choice in Rats as a Function of Reinforcer Intensity and Quality: A Thesis Presented to the Faculty of California State University, Fullerton, in Partial Fulfillment of the Requirements for the Degree Master of Arts in Psychology," Douglas J. Navarick, committee chair; Margaret H. White and Michael J. Scavio, members, August 8, 1978.

2. Richard J. Herrnstein, "Relative and Absolute Strength of Response as a Function of Frequency of Reinforcement," *Journal of the Experimental Analysis of Behavior* 4 (1961): 267–72.

3. Peter A. De Villiers and Richard Herrnstein, "Toward a Law of Response Strength," *Psychological Bulletin* 83 (1976): 1131–53.

4. Thomas Gilovich and Gary Belsky, *Why Smart People Make Big Money Mistakes and How to Correct Them: Lessons from the New Science of Behavioral Economics* (New York: Fireside, 2000).

5. Daniel Kahneman and Amos Tversky, "Prospect Theory: An Analysis of Decision Under Risk," *Econometrica* 47(2) (1979): 263–91. Reprinted in Daniel Kahneman and Amos Tversky, *Choices, Values, and Frames* (New York: Cambridge University Press, 2000), 17–43. What Kahneman and Tversky did for economics was to integrate the research findings and methodologies from their own field of psychology into the study of how people behave in markets. They embodied what an earlier integrationist, the Nobel laureate economist Friedrich Hayek, recommended for students of the profession: "An economist who is only an economist cannot be a good economist," a position quoted and endorsed by another Nobel economist, Vernon Smith, in his prize acceptance speech. Such transdisciplinary links between economics and psychology have proven to be some of the most fruitful in the history of both fields.

6. John Nash, "Equilibrium Points in N-Person Games," *Proceedings of the National Academy of Sciences* 36(1) (1950): 48–49; and "Non-Cooperative Games," *The Annals of Mathematics* 54(2) (1951): 286–95; Oskar Morgenstern and John von Neumann, *The Theory of Games and Economic Behavior* (Princeton, NJ: Princeton University Press, 1947).

7. For a general discussion on the most recent research on game theory in light of such concepts as Nash equilibrium and Pareto optimization, see Colin F. Camerer, *Behavioral Game Theory: Experiments in Strategic Interaction* (Princeton, NJ: Princeton University Press, 2003).

8. Colin Camerer et al., "Labor Supply of New York City Cabdrivers: One Day at a Time," *The Quarterly Journal of Economics* 112(2) (May 1997): 407–41. Camerer's career has been dedicated to this methodological triad ever since he came to Caltech in 1994 and opened his laboratory in behavioral economics, where his star is well ensconced among the galaxy of talent clustered on this small Pasadena campus. Ambition runs high here. Behavioral economists want to use their science to inform policy with an end goal of determining how we can improve people's and society's welfare so that everyone is better off and no one is worse off. Ambitious indeed.

9. Keith Chen, Venkat Lakshminarayanan, and Laurie Santos, "How Basic are Behavioral Biases? Evidence from Capuchin-Monkey Trading Behavior," *Journal of Political Economy*, June 2006.

10. Richard H. Thaler, "Some Empirical Evidence on Dynamic Inconsistency," *Economic Letters* 8 (1981): 201–7.

11. H. Rachlin, *Judgment, Decision and Choice: A Cognitive/Behavioral Synthesis* (New York: W. H. Freeman, 1989), chap. 7; J. H. Kagel, R. C. Battalio, and L. Green, *Economic Choice Theory: An Experimental Analysis of Animal Behavior* (Cambridge: Cambridge University Press, 1995).

12. R. Antonio Damasio, *Descartes' Error: Emotion, Reason, and the Human Brain* (New York: G. P. Putnam, 1994); Joseph E. Ledoux, *The Emotional Brain: The Mysterious Underpinnings of Emotional Life* (New York: Simon & Schuster, 1996).

13. Samuel M. McClure, David I. Laibson, George Loewenstein, and Jonathan D. Cohen, "Separate Neural Systems Value Immediate and Delayed Monetary Rewards," *Science* 306 (2004): 503–7. Note: this interpretation of the limbic system has come into some disrepute among some neuroscientists, who argue that it is not clear that the different regions noted as part of this system are really functionally coherent. Also, recent work by Joe Kable and Paul Glimcher challenges the McClure interpretation, showing that activity in the limbic system regions decrease exactly as predicted by the subject's behavioral discounting function, and that their data are fit better by a single discounting function than by the beta/delta model suggested by McClure.

14. The amount of data generated in a single experiment is mind-boggling: a brain snapshot every two seconds for 30 minutes results in 900 photographs per subject at a total size of about 250 megabytes, in an experiment in which there could be as many as 16 or 20 subjects.

15. Sabrina Tom, Craig R. Fox, Christopher Trepel, and Russell A. Poldrack, "The Neural Basis of Loss Aversion in Decision-Making Under Risk," *Science* 315 (January 26, 2007): 515–18.

16. Interview with Russell Poldrack and Craig Fox conducted on March 12, 2007.

17. Daniel Kahneman, B. L. Fredrickson, C. A. Schreiber, and D. A. Redelmeier, "When More Pain Is Preferred to Less: Adding a Better End," *Psychological Science* 4 (1993): 401–5.

18. Michael L. Platt and Paul W. Glimcher, "Neural Correlates of Decision Variables in Parietal Cortex," *Nature* 400 (July 15, 1999): 234. See also Paul W. Glimcher, *Decisions, Uncertainty, and the Brain* (Cambridge, MA: MIT Press, 2003).

19. B. Knutson, S. Rick, G. E. Wimmer, D. Prelec, and G. Loewenstein, "Neural Predictors of Purchases," *Neuron* 53 (2007): 147–57.

20. Quoted in "This Is Your Brain on Shopping," *Scientific American*, January 5, 2007, www.sciamdigital.com.

21. James Olds and Peter Milner, "Positive Reinforcement Produced by Electrical Stimulation of Septal Area and Other Regions of Rat Brain," *Journal of Comparative and Physiological Psychology* 47 (1954): 419–27.

22. M. E. Olds and J. L. Fobes, "The Central Basis of Motivation: Intracranial Self-Stimulation Studies," *Annual Review of Psychology* 32 (1981): 523–74; M. P. Bishop, S. T. Elder, and R. G. Heath, "Intracranial Self-Stimulation in Man," *Science* 140 (1963): 394–96.

23. C. M. Kuhnen and B. Knutson, "The Neural Basis of Financial Risk-Taking," *Neuron* 47 (2005): 768.

7. The Value of Virtue

1. The trolley car thought experiment was first proposed by the philosopher Phillipa Foot in "The Problem of Abortion and the Doctrine of Double Effect," *Oxford Review* 5 (1967): 5–15. The extensive research utilizing the trolley car scenario has been summarized in many works, most recently in Marc Hauser, *Moral Minds: How Nature Designed Our Universal Sense of Right and Wrong* (New York: HarperCollins, 2006), 33–34, 113–20. Hauser, who has conducted his own research on the trolley car dilemma (http://moral.wjh.harvard.edu), argues that the experimental results reveal that we have an evolved "moral grammar," not unlike our evolved language grammar. Just as we are born with a capacity to learn a language complete with rules of grammar that are fined-tuned by the culture in which we are raised, so too are we born with a capacity to be moral, the specific moral grammar rules of which are determined by the specific culture in which we are raised. See also L. Petrinovich, P. O'Neill, and M. J. Jorgensen, "An Empirical Study of Moral Intuitions: Towards an Evolutionary Ethics," *Ethology and Sociobiology* 64 (1993): 467–78.

2. A much longer essay on my experiences with my mom's cancer, entitled "Shadowlands," is reprinted in my book *Science Friction: Where the Known Meets the Unknown* (New York: Times Books, 2005).

3. Michael Shermer, *The Science of Good and Evil* (New York: Times Books, 2004). The case for a universal moral sense continues to grow as new research is conducted on nonhuman primates and a variety of social mammals, from which it is clear that social relations and conflict resolution between individuals is at least as important as other traits in determining survival.

4. David M. Buss, *The Evolution of Desire: Strategies of Human Mating* (New York: Basic Books, 2003); D. Singh, "Adaptive Significance of Female Attractiveness: Role of Waist-to-Hip Ratio," *Journal of Personality and Social Psychology* 65 (1993): 293–307; Helen Fisher, *Why We Love: The Nature and Chemistry of Romantic Love* (New York: Henry Holt, 2004).

5. Extensive references for this research are provided in Shermer, *The Science of Good and Evil*, most notably chapters 2, 7, and 8. On one level, *The Mind of the Market* is a continuation of where I left off in chapter 8, the final chapter of that book.

6. B. J. Ellis, "The Evolution of Sexual Attraction: Evaluative Mechanisms in Women," in J. H. Barkow, L. Cosmides, and J. Tooby, eds., *The Adapted Mind: Evolutionary Psychology and the Generation of Culture* (New York: Oxford UP, 1992), 267–88; Buss, *Evolution of Desire*; T. Bereczkei and A. Csanaky, "Mate Choice, Marital Success, and Reproduction in a Modern Society," *Ethology & Sociobiology* 17(1) (1996): 17–35; Randy Thornhill and S. W. Gaugestad, "Fluctuating Asymmetry and Human Sexual Behavior," *Psychological Science* 5 (1994): 297–302.

7. Donald E. Brown, *Human Universals* (New York: McGraw-Hill, 1991), 142.

8. Arthur Gandolfi, Anna Sachko Gandolfi, and David Barash, *Economics as an Evolutionary Science* (New Brunswick, NJ: Transaction Publishers, 2002), 139–40.

9. Study cited and discussed in David Buss, *The Dangerous Passion: Why Jealousy Is as Necessary as Love and Sex* (New York: The Free Press, 2000). The expression "zipless fuck" comes from Erica Jong's 1973 novel *Fear of Flying*: "The zipless fuck is absolutely pure. It is free of ulterior motives. There is no power game. The man is not 'taking' and the woman is not 'giving.' No one is attempting to cuckold a husband or humiliate a wife. No one is trying to prove anything or get anything out of anyone. The zipless fuck is the purest thing there is. And it is rarer than the unicorn."

10. Gandolfi et al., *Economics as an Evolutionary Science*, 139–89. I added the modifier "relatively" since, as we all know, plenty of men are faithful to their wives, and plenty of women have affairs. Reliable data to define "plenty" is hard to come by, and published figures range widely, but surveys typically find that between 25 and 50 percent of U.S. men and around 30 percent of women report having had at least one episode of extramarital sex. See David P. Barash and Judith Eve Lipton, *The Myth of Monogamy: Fidelity and Infidelity in Animals and People* (New York: W. H. Freeman, 2001).

11. S. W. Gangestad and J. A. Simpson, "The Evolution of Human Mating: Trade-Offs and Strategic Pluralism," *Behavioral and Brain Sciences* 23(4) (2000): 1–33.

12. Martin Daly and Margo Wilson, "Child Abuse and Other Risks of Not Living with Both Parents," *Ethology & Sociobiology* 6 (1985): 197–210; Daly and Wilson, *Homicide* (New York: Aldene de Gruyter, 1988); Daly and Wilson, "Stepparenthood and the Evolved Psychology of Discriminative Parental Solicitude," in S. Parmigiani and F. S. Von Saal, eds., *Infanticide and Parental Care* (London: Harwood Press, 1994), 121–34.

13. L. K. White and A. Booth, "The Quality and Stability of Remarriage: The Role of Step-Children," *American Sociological Review* 50 (1985): 346–58; M. V. Flinn, "Step- and Genetic Parent/Offspring Relationships in a Caribbean Village," *Ethology & Sociobiology* 9 (1988): 335–69.

14. D. A. Dawson, "Family Structure and Children's Health and Well-being: Data from the 1988 National Health Interview Survey on Child Health," *Journal of Marriage and Family* 53 (1991): 573–84; K. E. Kierman, "The Impact of Family Disruption in Childhood on Transitions Made in Young Adult Life," *Population Studies* 46 (1992): 213–34.

15. M. Gordon, "The Family Environment of Sexual Abuse: A Comparison of Natal and Step-Father Abuse," *Child Abuse and Neglect* 13 (1989): 121–30; D. E. H. Russel, "The Prevalence and Seriousness of Incestuous Abuse: Stepfathers vs Biological Fathers," *Child Abuse and Neglect* 8 (1984): 15–22; M. Young, *The Sexual Victimization of Children* (Jefferson, NC: McFarland Press, 1982).

16. M. K. Bacon, I. L. Child, and H. Bary, "A Cross-Cultural Study of Correlates of Crime," *Journal of Abnormal and Social Psychology* 66 (1963): 291–300.

17. Daly and Wilson, *Homicide*.

18. Margo Wilson and Martin Daly, "Competitiveness, Risk Taking and Violence: The Young Male Syndrome," *Ethology and Sociobiology* 6 (1985): 59–73.

19. Steven Pinker, *The Blank Slate: The Modern Denial of Human Nature* (New York: Viking, 2002).

20. Gary Becker, *A Treatise on the Family* (Cambridge, MA: Harvard University Press, 1981); Laura Betzig, *Despotism and Human Reproduction: A Darwinian Viewpoint* (New York: Aldine, 1986).

21. Margo Wilson and Martin Daly, "The Man Who Mistook His Wife for a Chattel," in Jerome Barko, Leda Cosmides, and John Tooby, eds., *The Adapted Mind* (New York: Oxford University Press, 1992), 289–322.

22. Daly and Wilson, *Homicide*; Daly and Wilson, "Violence Against Stepchildren," *Current Directions in Psychological Science* 5 (1996): 77–81.

23. Daly and Wilson, "Whom Are Newborn Babies Said to Resemble?" *Ethology and Sociobiology* 3 (1982): 69–78.

24. Jared Diamond, *Why Is Sex Fun? The Evolution of Human Sexuality* (New York: Basic Books, 1997), 76.

25. Shermer, *The Science of Good and Evil*, chap. 4, "Master of My Fate," 105–38.

26. Richard D. Alexander, *Darwinism and Human Affairs* (Seattle: University of Washington Press, 1979); and *The Biology of Moral Systems* (New York: Aldine De Gruyter, 1987); Frank Miele, "The (Im)moral Animal: A Quick and Dirty Guide to Evolutionary Psychology and the Nature of Human Nature," *Skeptic* 4(1) (1996): 42–49.

27. Edward O. Wilson, *Biophilia* (Cambridge, MA: Harvard University Press, 1984).

28. Chris Boehm, "Egalitarian Society and Reverse Dominance Hierarchy," *Current Anthropology* 34 (1993): 227–54; and *Hierarchy in the Forest: Egalitarianism and the Evolution of Human Altruism* (Cambridge, MA: Harvard University Press, 1999).

29. David Sloan Wilson, *Darwin's Cathedral: Evolution, Religion, and the Nature of Society* (Chicago: University of Chicago Press, 2002).

30. Signe Howell, *Society and Cosmos: Chewong of Peninsular Malaya* (Singapore: Oxford University Press, 1984), 184.

31. Amotz Zahavi and Avishag Zahavi, *The Handicap Principle: A Missing Piece of Darwin's Puzzle* (Oxford: Oxford University Press, 1997).

32. R. Adolphs, "Cognitive Neuroscience of Human Social Behavior," *Nature Reviews Neuroscience* 4 (March 2003): 165–70; R. J. Dolan, "Emotion, Cognition, and Behavior," *Science* 298 (November 8, 2002): 1191–94.

33. K. McCabe, D. Houser, L. Ryan, V. Smith, and T. Trouard, "A Functional Imaging Study of Cooperation in Two-Person Reciprocal Exchange," *Proceedings of the National Academy of Science* 98(20) (2001): 11832–35.

34. K. Semendeferi, E. Armstrong, A. Schleicher, K. Zilles, and G. W. van Hoesen, "Prefrontal Cortex in Humans and Apes: A Comparative Study of Area 10," *American Journal of Physical Anthropology* 114 (2001): 224–41.

35. J. Moll, R. de Oliveira-Souza, P. J. Eslinger, I. E. Bramati, J. Mourai-Miranda, P. A. Andreiuolo, and L. Pessoa, "The Neural Correlates of Moral Sensitivity: A Functional Magnetic Resonance Imaging Investigation of Basic and Moral Emotions," *The Journal of Neuroscience* 22(7) (April 1, 2002): 2730–36.

36. Jorge Moll, Frank Krueger, Roland Zahn, Matteo Pardini, Ricardo de Oliveira-Souza, and Jordan Grafman, "Human Fronto-Mesolimbic Networks Guide Decisions About Charitable Donation," *Proceedings of the National Academy of Science*, 103(42) (2006): 15623–28.

37. U. Frith and C. Frith, "The Biological Basis of Social Interaction," *Current Directions in Psychological Science* 10(5) (2001): 151–55.

38. Helen Phillips, "The Cell That Makes Us Human," *New Scientist*, June 2004, 33–35; J. M. Allman, K. K. Watson, N. A. Tetreault, and A. Y. Hakeem, "Intuition and Autism: A Possible Role for Von Economo Neurons," *Trends in Cognitive Sciences* 9(8) (2005): 367–73; K. K. Watson, T. K. Jones, and J. M. Allman, "Dendritic Architecture of the Von Economo Neurons," *Neuroscience* 141 (2006): 1107–12.

39. Giacomo Rizzolatti et al., "Premotor Cortex and the Recognition of Motor Actions," *Cognitive Brain Research* 3 (1996): 131–41.

40. L. Fogassi, P. F. Ferrari, B. Gesierich, S. Rozzi, F. Chersi, and G. Rizzolatti, "Parietal Lobe: From Action Organization to Intention Understanding," *Science* 308 (2005): 662–67; V. Gallese, L. Fadiga, L. Fogassi, and G. Rizzolatti, "Action Recognition in the Premotor Cortex," *Brain* 119 (1996), 593–609.

41. M. Iacoboni, R. P. Woods, M. Brass, H. Bekkering, J. C. Mazziotta, and G. Rizzolatti, "Cortical Mechanisms of Human Imitation," *Science* 286 (1999): 2526–28; G. Rizzolatti and L. Craighero, "The Mirror-Neuron System," *Annual Review of Neuroscience* 27 (2004): 169–92. It should be noted that the activity imaged in such fMRI studies is not the same as the recording of individual neurons in monkeys' brains. As the University of Gröningen psychologist Christian Keysers explained, "When we record signals from neurons in monkeys, we can really know that a single neuron is involved in both doing the task and seeing someone else do the task. With imaging, you know that within a little box about three millimeters by three millimeters by three millimeters, you have activation from both doing and seeing. But this little box contains millions of neurons, so you cannot know for sure that they are the same neurons—perhaps they're just neighbors." See Lea Winerman, "The Mind's Mirror," *Monitor on Psychology* 36(9) (October 2005): 48.

42. G. Buccino, S. Vogt, A. Ritzl, G. R. Fink, K. Zilles, H. J. Freund, and G. Rizzolatti, "Neural Circuits Underlying Imitation of Hand Actions: An Event Related fMRI Study," *Neuron* 42 (2004): 323–34.

43. Helen L. Gallagher and Christopher D. Frith, "Functional Imaging of 'Theory of Mind,'" *Trends in Cognitive Sciences* 7(2) (February 2003): 77.

44. V. Gallese and A. Goldman, "Mirror Neurons and the Simulation Theory of Mind-Reading," *Trends in Cognitive Sciences* 12 (1998): 493–501.

45. L. Fogassi, P. F. Ferrari, B. Gesierich, S. Rozzi, F. Chersi, and G. Rizzolatti, "Parietal Lobe: From Action Organization to Intention Understanding," *Science* 308 (2005): 662–67.

46. B. Wicker, C. Keysers, J. Plailly, J. P. Royet, V. Gallese, and G. Rizzolatti, "Both of Us Disgusted in My Insula: The Common Neural Basis of Seeing and Feeling Disgust," *Neuron* 40 (2003): 655–64.

47. L. Carr, M. Iacoboni, M. C. Dubeau, J. C. Mazziotta, and G. L. Lenzi, "Neural Mechanisms of Empathy in Humans: A Relay from Neural Systems for Imitation to Limbic Areas," *Proceedings of the National Academy of Science* 100 (2003): 5497–5502.

48. M. Iacoboni, I. Molnar-Szakacs, V. Gallese, G. Buccino, J. C. Mazziotta, et al., "Grasping the Intentions of Others with One's Own Mirror Neuron System," *PLoS Biology* 3(3) (2005): e79 doi:10.1371/journal.pbio.0030079.

49. Y. W. Cheng, O. J. L. Tzeng, J. Decety, T. Imada, and J. C. Hsieh, "Gender Differences in the Human Mirror System: A Magnetoencephalography Study," *Neuroreport* 17(11) (2006): 1115–19; Simon Baron-Cohen, *The Essential Difference: The Truth about the Male and Female Brain* (New York: Basic Books, 2003).

50. M. V. Saarela, Y. Hlushchuk, A. C. Williams, M. Schurmann, E. Kalso, and R. Hari, "The Compassionate Brain: Humans Detect Intensity of Pain from Another's Face," *Cerebral Cortex* (2006); T. Singer, "The Neuronal Basis and Ontogeny of Empathy and Mind Reading: Review of Literature and Implications for Future Research," *Neuroscience and Biobehavioral Reviews* 6 (2006): 855–63.

51. V. S. Ramachandran and L. M. Oberman, "Broken Mirrors: A Theory of Autism," *Scientific American*, May 2006, 62–69. For the connection between mirror neurons and the evolution of language, see M. A. Arbib, "From Monkey-Like Action Recognition to Human Language: An Evolutionary Framework for Neurolinguistics," *The Behavioral and Brain Sciences* 2 (2005): 105–24.

52. V. S. Ramachandran, "Mirror Neurons and Imitation Learning as the Driving Force Behind 'The Great Leap Forward' in Human Evolution," *Edge* (2000), available at http://www.edge.org/3rd_culture/ramachandran/ramachandran_index.html.

53. Adam Smith, *The Theory of Moral Sentiments* (London: A. Millar, 1759), part 1, sec. 1, chap. 1. The last two decades have witnessed the development of a number of theories on the evolutionary origins of the moral emotions. See, for example, William D. Casebeer, "Moral Cognition and Its Neural Constituents," *Nature Neuroscience* 4 (2003): 841–46; William D. Casebeer and Patricia S. Churchland, "The Neural Mechanisms of Moral Cognition: A Multiple-Aspect Approach to Moral Judgment and Decision-Making," *Biology and Philosophy* 18 (2003): 169–94; Jorge Moll, Roland Zahn, Ricardo de Oliveira-Souza, Frank Krueger, and Jordan Grafman, "The Neural Basis of Human Moral Cognition," *Nature Neuroscience* 6 (2005): 799–809; Richard Dawkins, *The God Delusion* (New York: Houghton Mifflin, 2006); Marc Hauser, *Moral Minds: How Nature Designed Our Universal Sense of Right and Wrong* (New York: HarperCollins, 2006); Scott Atran, *In Gods We Trust: The Evolutionary Landscape of Religion* (New York: Oxford University Press, 2004); Robert Hinde, *Why Good Is Good: The Sources of Morality* (New York: Routledge, 2003); Pascal Boyer, *Religion Explained: The Evolutionary Origins of Religious Thought* (New York: Basic Books, 2001); and Robert Buckman, *Can We Be Good Without God?* (Buffalo, NY: Prometheus, 2000).

54. The final scene from *About Schmidt* is available on YouTube by searching "Ndugu's Letter." Have some tissues handy.

55. I thank my friend Robbin Gehrke, who works for the public relations firm that represents World Vision, for telling me about this sponsorship program and its ability to deliver support reliably and efficiently where most needed.

8. Why Money Can't Buy You Happiness

1. Jeremy Bentham, *The Principles of Morals and Legislation* (New York: Macmillan, 1948) (orig. pub. 1789).

2. From Penn World Tables. Real GDP per capita (Laspeyres) (RGDPL) is obtained by adding up consumption, investment, government spending, and exports, and subtracting imports in any given year. The given year components are obtained by extrapolating the 1996 values in international dollars from the Geary aggregation using national growth rates. It is a fixed base index where the reference year is 1996, hence the designation "L" for Laspeyres. Computed by Levan Efremidze.

3. These economic measures are summarized thoroughly and in great detail by Easterbrook (see following note) and briefly in Richard Layard, *Happiness: Lessons from a New Science* (New York: Penguin, 2005).

4. David G. Myers, *The American Paradox: Spiritual Hunger in an Age of Plenty* (New Haven, CT: Yale University Press, 2000); Robert E. Lane, *The Loss of Happiness in Market Democracies* (New Haven, CT: Yale University Press, 2000); Gregg Easterbrook, *The Progress Paradox: How Life Gets Better While People Feel Worse* (New York: Random House, 2003); Barry Schwartz, *The Paradox of Choice: Why More Is Less* (New York: Ecco/HarperCollins, 2004).

5. W. Pavot and E. Diener, "Review of the Satisfaction with Life Scale," *Psychological Assessment* 5 (1993): 164–72.

6. E. Diener and E. Suh, "National Differences in Subjective Well-Being," in Daniel Kahneman, E. Diener, and N. Schwarz, eds., *Well-Being: The Foundations of Hedonic Psychology* (New York: Russell Sage Foundation, 1999).

7. Joel M. Hektner, Jennifer A. Schmidt, and Mihaly Csikszentmihalyi, *Experience Sampling Method: Measuring the Quality of Everyday Life* (Thousand Oaks, CA: Sage, 2006). The Experience Sampling Method involves giving subjects beepers and a survey

instrument measuring subjective well-being, and then beeping them randomly throughout the week, at which point they complete the survey questions. This avoids the problem of self-report happiness data being time- and context-dependent.

8. Study conducted by the Princeton Research Associates, July 12–15, cited in *American Enterprise*, November/December 1994, 99.

9. Rafael di Tella, Robert J. MacCulloch, and Andrew J. Oswald, "The Macroeconomics of Happiness," Warwick Economic Research Papers no. 615, 2001.

10. World Values Survey, http://www.worldvaluessurvey.com/.

11. R. Inglehart and H.-D. Klingemann, "Genes, Culture, Democracy and Happiness," in E. Diener and E. M. Suh, eds., *Culture and Subjective Well-Being* (Cambridge, MA: MIT Press, 2000).

12. N. G. Martin, L. J. Eaves, A. C. Heath, R. Jardine, L. M. Feingold, and H. J. Eysenck, "Transmission of Social Attitudes," *Proceedings of the National Academy of Science USA* 83 (1986): 4364–68; L. J. Eaves, H. J. Eysenck, and N. G. Martin, *Genes, Culture and Personality: An Empirical Approach* (London and San Diego: Academic Press, 1989).

13. N. G. Waller, B. Kojetin, T. Bouchard, D. Lykken, and A. Tellegen, "Genetic and Environmental Influences on Religious Attitudes and Values: A Study of Twins Reared Apart and Together," *Psychological Science* 1(2) (1990): 138–42.

14. David T. Lykken, *Happiness: The Nature and Nurture of Joy and Contentment* (New York: St. Martin's Griffin, 2000).

15. Sara Solnick and David Hemenway, "Is More Always Better? A Survey on Positional Concerns," *Journal of Economic Behavior and Organization*, 37 (1998): 373–83.

16. Fredrik Carlsson, Olof Johansson-Stenman, and Peter Martinsson, "Do You Enjoy Having More than Others? Survey Evidence of Positional Goods," *Economica* (Online Early Articles), 2007, http://www.blackwell-synergy.com/doi/full/10.1111/j.1468-0335.2006.00571.x.

17. B. Van Praag and P. Frijters, "The Measurement of Welfare and Well-Being: The Leyden Approach," in Kahneman, Diener, and Schwarz, *Well-Being*. An earlier economist to notice this effect was Tibor Scitovsky, in his *The Joyless Economy: An Inquiry into Human Satisfaction and Consumer Dissatisfaction* (New York: Oxford University Press, 1978).

18. Richard Thaler, "Toward a Positive Theory of Consumer Choice," *Journal of Economic Behavior and Organization*, 1980, reprinted in Breit and Hochman, eds., *Readings in Microeconomics*, 3rd ed., 1985; and in Howard Kunreuther, ed., *Risk: A Seminar Series*, IIASA, 1981.

19. Gregory Berns, *Satisfaction* (New York: Owl Books, 2005).

20. Daniel T. Gilbert, Elizabeth Pinel, Timothy Wilson, Stephen Blumberg, and Thalia Wheatley, "Immune Neglect: A Source of Durability Bias in Affective Forecasting," *Journal of Personality and Social Psychology* 75 (1998): 633.

21. Daniel Gilbert, *Stumbling on Happiness* (New York: Knopf, 2005).

22. E. O. Laumann, J. H. Gagnon, R. T. Michael, and S. Michaels, *The Social Organization of Sexuality: Sexual Practices in the United States* (Chicago: University of Chicago Press, 1994).

23. Jennifer Hecht, *The Happiness Myth: Why Smarter, Healthier, and Faster Doesn't Work* (San Francisco: Harper, 2007), 223–25.

24. David G. Myers, *The Pursuit of Happiness: Who Is Happy and Why* (New York: William Morrow, 1992); Martin E. P. Seligman, *Authentic Happiness* (New York: Free Press, 2002).

25. D. A Schade, and Daniel Kahneman, "Does Living in California Make People Happy? A Focusing Illusion in Judgments of Life Satisfaction," *Psychological Science* 9(5) (1998): 329–40.

26. Elizabeth W. Dunn, T. D. Wilson, and Daniel T. Gilbert, "Location, Location, Location: The Misprediction of Satisfaction in Housing Lotteries," *Personality and Social Psychology Bulletin* 29(11) (2003): 1421–32.

27. Richard Davidson, "Emotion and Affective Style: Hemispheric Substrates," *Psychological Science* 3 (1992): 39–43.

28. N. Fox and Richard Davidson, "Taste-Elicited Changes in Facial Signs of Emotion and the Asymmetry of Brain Electrical Activity in Human Newborns," *Neuropsychologia* 24 (1986): 417–22.

29. S. Lisanby, "Focal Brain Stimulation with Repetitive Transcranial Magnet Stimulation (rTMS): Implications for the Neural Circuitry of Depression," *Psychological Medicine* 33 (2003): 7–13.

30. A. Quaranta, M. Siniscalchi, and G. Vallortigara, "Asymmetric Tail-Wagging Responses by Dogs to Different Emotive Stimuli," *Current Biology* 17 (2007): R199–R201.

31. Reported in Layard, *Happiness*, 19.

32. Lykken, *Happiness*.

33. Antonio R. Damasio, *Descartes' Error: Emotion, Reason, and the Human Brain* (New York: Putnam, 1994); Ellen Peters and Paul Slovic, "The Springs of Action: Affective and Analytical Information Processing in Choice," *Personality and Social Psychological Bulletin* 26(12) (2000): 1465–75; Jon Elster, *Ulysses and the Sirens* (Cambridge: Cambridge University Press, 1977); Roy F. Baumeister, Todd F. Heatherton, and Dianne M. Tice, *Losing Control: How and Why People Fail at Self-Regulation* (San Diego: Academic Press, 1994); George Loewenstein, "Out of Control: Visceral Influences on Behavior," *Organizational Behavior and Human Decision Processes* 65 (1996): 272–92; George F. Loewenstein and Jennifer Lerner, "The Role of Affect in Decision Making," in *Handbook of Affective Sciences*, R. J. Davidson, K. R. Scherer, and H. H. Goldsmith, eds. (New York: Oxford University Press, 1996), 619–42.

34. Björn Grinde, *Darwinian Happiness: Evolution as a Guide for Living and Understanding Human Behavior* (Princeton, NJ: Darwin Press, 2002), 49. See also his article "Happiness in the Perspective of Evolutionary Psychology," *Journal of Happiness Studies* 3 (2002): 331–54.

35. Peggy La Cerra and Roger Bingham, *The Origin of Minds: Evolution, Uniqueness, and the New Science of the Self* (New York: Harmony Books, 2003).

36. Jennifer Lerner and Dacher Keltner, "Fear, Anger, and Risk," *Journal of Personality and Social Psychology* 81(2001): 146–59.

37. Colin Camerer, George Loewenstein, and Drazen Prelec, "Neuroeconomics: How Neuroscience Can Inform Economics," *Journal of Economic Literature* 34(1) (2005): 9–65. See also Andrew Oswald, "Happiness and Economic Performance," *Economic Journal* 107 (1997): 1815–31; and Wolfram Schulz and A. Dickinson, "Neuronal Coding of Prediction Errors," *Annual Review of Neuroscience* 23 (2000): 473–500.

38. T. D. Wilson, D. J. Lisle, J. W. Schooler, S. D. Hodges, K. J. Klaaren, and S. J. LaFleur, "Introspecting About Reasons Can Reduce Post-Choice Satisfaction," *Personality and Social Psychology Bulletin* 19 (1993): 331–39; Timothy D. Wilson and Jonathan W. Schooler, "Thinking Too Much: Introspection Can Reduce the Quality of Preferences and Decisions," *Journal of Personality and Social Psychology* 60(2) (1991): 181–92.

39. K. C. Berridge, "Food Reward: Brain Substrates of Wanting and Liking," *Neuroscience and Biobehavioral Reviews* 20 (1996): 1–25.

40. Personal correspondence, April 10, 2007.

41. Barbara L. Fredrickson, "The Value of Positive Emotions," *American Scientist* 91, (July/August 2003): 330–35.

42. Gordon M. Burghardt, *The Genesis of Animal Play* (Cambridge, MA: MIT Press, 2005), 113–17.

43. Daniel Kahneman, A. Krueger, D. Schkade, N. Schwarz, and A. Stone, "A Survey Method for Characterizing Daily Life Experiences: The Day Reconstruction Method (DRM)," *Science* 306 (2004): 1776–80.

44. David G. Myers and Ed Diener, "The Pursuit of Happiness," *Scientific American*, May 1996, 70–72.

45. Ibid., 72. See also David G. Myers, *The Pursuit of Happiness: Who Is Happy and Why* (New York: William Morrow, 1992).

46. Michael Shermer, *How We Believe: The Search for God in an Age of Science* (New York: Times Books, 1999).

47. D. B. Barrett, G. T. Kurian, and T. M. Johnson, eds., *World Christian Encyclopedia: A Comparative Survey of Churches and Religions in the Modern World*, 2 vols., 2nd ed. (New York: Oxford University Press, 2001). Christians dominate at just a shade under 2 billion adherents (Catholics count for half of those), with Muslims at 1.1 billion, Hindus at 811 million, Buddhists at 359 million, and ethnoreligionists (animists and others in Asia and Africa primarily) accounting for most of the remaining 265 million.

48. Interview of Mihaly Csikszentmihalyi conducted by the author on April 17, 2007. See also Mihaly Csikszentmihalyi, *Flow: The Psychology of Optimal Experience* (New York: HarperCollins, 1991); and *Finding Flow: The Psychology of Engagement with Everyday Life* (New York: Basic Books, 1998).

49. Helen Keller, "The Simplest Way to Be Happy," *Home Magazine*, February 1993, available at the Web site of the American Foundation for the Blind: http://www.afb.org/section.asp?SectionID=1&TopicID=193&SubTopicID=12&DocumentID=1211.

9. Trust with Credit Verification

1. The story appears in Patrick Grim's teaching company course, *Questions of Value*, Lecture 23, "Moralities in Conflict and Change." Although I am using the story to make a different point than Grim does, the deeper meaning of it is the same.

2. Interview conducted in November 2000. I thank my friends Donna and Michael Coles for making this visit to De Waal's lab possible. The full interview was originally published in *Skeptic*, vol. 8, no. 2.

3. I also asked De Waal about the policy implications of research on the evolutionary basis of moral behavior. For example, as we saw in chapter 7, most crimes are committed by young men in their twenties, in part because of high levels of testosterone coupled with the fact that as social hierarchical primates, young men go through initiation rites, compete for females, struggle for power with other males, etc., all of which can lead to antisocial behavior. How can evolutionary theory inform our political debates about moral issues? Here De Waal once again employs his skepticism about pushing evolutionary models too far. Evolutionary theory, he thinks, "can help us understand why we have moral systems and how they operate. But I would never say 'that's how chimps do it so we should too.' First of all, even though we are so closely related to chimps, we have rather different kinds of societies. Even within human groups, societies vary considerably, along with their moral systems." How, then, should we determine how to act morally? "The decision on how we should act is a consensual decision we make as a society. Now, in that decision enters human nature—we have tendencies of sympathy, reciprocity, parent-child relationships, aggression, or whatever. For example, if you set up a society that is in direct conflict with one of these basic

human needs you are setting yourself up for failure." Here I think of the Israeli kibbutz where children were separated from their parents and raised by the group, but it was a failed social experiment because mothers still wanted to be with their own children.

4. Frans de Waal, *Chimpanzee Politics: Sex and Power Among the Apes* (Baltimore: Johns Hopkins University Press, 1982), 203, 207.

5. Frans de Waal, *Peacemaking Among Primates* (Cambridge, MA: Harvard University Press, 1989).

6. Frans de Waal, *Our Inner Ape* (New York: Riverhead Books, 2005), 175.

7. Ibid., 187–88.

8. Frans B. M. de Waal, "Food-transfers Through Mesh in Brown Capuchins," *Journal of Comparative Psychology* 111 (1997): 370–78; and "Food Sharing and Reciprocal Obligations Among Chimpanzees," *Journal of Human Evolution* (1989): 433–59.

9. Sarah F. Brosnan and Frans de Waal, "Monkeys Reject Unequal Pay," *Nature* 425 (September 18, 2003): 297–99.

10. M. Cords and S. Thurnheer, "Reconciling with Valuable Partners by Long-Tailed Macaques," *Behaviour* 93 (1993): 315–25.

11. Sarah F. Brosnan, "Fairness and Other-Regarding Preferences in Nonhuman Primates," in Paul Zak, ed., *Moral Markets: The Critical Role of Values in the Economy* (Princeton, NJ: Princeton University Press, 2008).

12. Frans de Waal, *Good Natured: The Origins of Right and Wrong in Humans and Other Animals* (Cambridge, MA: Harvard University Press, 1996).

13. N. F. Koyama and E. Palagi, "Managing Conflict: Evidence from Wild and Captive Primates," *International Journal of Primatology* 27(5) (2007): 1235–40; N. F. Koyama, C. Caws, and F. Aureli, "Interchange of Grooming and Agonistic Support in Chimpanzees," *International Journal of Primatology* 27(5) (2007): 1293–1309.

14. Daniel Dennett, "True Believers" in D. Dennett, ed., *The Intentional Stance* (Cambridge, MA: MIT Press, 1987); and *Breaking the Spell: Religion as a Natural Phenomenon* (New York: Penguin, 2005).

15. Frans de Waal, "How Selfish an Animal? The Case of Primate Cooperation," in Zak, ed., *Moral Markets*.

16. Interviews conducted over a series of visits in February and March 2007.

17. Paul J. Zak, "Trust," *Capco Institute Journal of Financial Transformation* 7 (2003): 13–21.

18. Paul J. Zak and Stephen Knack, "Trust and Growth," *The Economic Journal* 111 (2001): 295–321.

19. Zak, "Trust."

20. V. C. L. Hutson and G. T. Vickers, "The Spatial Struggle of Tit-for-Tat and Defect," *Philosophical Transactions of the Royal Society of London B* 348 (1995): 393–404; K. Binmore, *Game Theory and the Social Contract*, vol. 1: *Playing Fair* (Cambridge, MA: MIT Press, 1994).

21. There is a sizable body of literature on game theory and cooperation. See, for example, John Von Neumann and Oskar Morgenstern, *Theory of Games and Economic Behavior* (Princeton, NJ: Princeton University Press, 2004 [orig. pub. 1944]); Robert L. Trivers, "The Evolution of Reciprocal Altruism," *Quarterly Review of Biology* 46 (1971): 35–57; Robert Axelrod and William D. Hamilton, "The Evolution of Cooperation," *Science* 211 (1981): 1390–96; Robert Axelrod, *The Evolution of Cooperation* (New York: Basic Books, 1984); Douglas R. Hofstadter, "Metamagical Themas: Computer Tournaments of the Prisoner's Dilemma Suggest How Cooperation Evolves," *Scientific American* 248(5) (1983): 16–26; Robert Frank, *Passions Within Reason: The Strategic Role of*

the Emotions (New York: W.W. Norton, 1988); Matt Ridley, *The Origins of Virtue* (New York: Viking, 1996); J. K. Murnighan, *Bargaining Games* (New York: William Morrow, 1992); M. Taylor, *The Possibility of Cooperation* (Cambridge: Cambridge University Press, 1987); and on the Internet, a Google search on "prisoner's dilemma" will lead to thousands of sites, computer simulations, chat rooms, discussions, bibliographies, etc., such as http://pespmc1.vub.ac.be/PRISDIL.html.

22. B. Bower, "Getting Out from Number One: Selfishness May Not Dominate Human Behavior," *Science News* 137(17) (1990): 266–67.

23. R. M. Dawes, A. van de Kragt, and J. M. Orbell, "Cooperation for the Benefit of Us—Not Me, or My Conscience," in J. Mansbridge, ed., *Beyond Self-Interest* (Chicago: University of Chicago Press, 1990), 97–110.

24. James Rilling, D. A. Gutman, T. R. Zeh, G. Pagnoni, G. S. Berns, and C. D. Kilts, "A Neural Basis for Social Cooperation," *Neuron* 35 (July 18, 2002): 395–404.

25. R. Forsythe, J. L. Horowitz, N. E. Savin, and M. Sefton, "Fairness in Simple Bargaining Experiments," *Games and Economic Behavior* 6 (1994): 347–69.

26. Helen Fisher, *Why We Love: The Nature and Chemistry of Romantic Love* (New York: Henry Holt, 2004).

27. Steven R. Quartz and Terry J. Sejnowski, *Liars, Lovers, and Heroes: What the New Brain Science Reveals About How We Become Who We Are* (New York: William Morrow, 2002).

28. M. Kosfeld, M. Heinrichs, P. J. Zak, U. Fischbacher, and E. Fehr, "Oxytocin Increases Trust in Humans," *Nature* 435 (2005): 673–76.

29. Paul J. Zak, "Trust: A Temporary Human Attachment Facilitated by Oxytocin," *Behavioral and Brain Sciences* 28(3) (2005): 368–69; Vera B. Morhenn, Jang Woo Park, Elisabeth Piper, and Paul J. Zak, "Monetary Sacrifice Among Strangers Is Mediated by Endogenous Oxytocin Release After Physical Contact," *Proceedings of the National Academy of Science* (in press).

30. Linda Mealy, "The Sociobiology of Sociopathy," *Behavioral and Brain Sciences* 18 (1995): 523–99; D. T. Lyyken, *The Antisocial Personalities* (Hillside, NJ: Lawrence Erlbaum Associates, 1995).

31. Ralph Adolphs, D. Tranel, and A. R. Damasio, "The Human Amygdala in Social Judgment," *Nature* 393 (1998): 470–74; Ralph Adolphs, D. Tranel, H. Damasio, and A. Damasio, "Impaired Recognition of Emotion in Facial Expressions Following Bilateral Damage to the Human Amygdala," *Nature* 372 (1994): 669–72.

32. Paul J. Zak, "Values and Value: Moral Economics," in Paul Zak, ed., *Moral Markets: The Critical Role of Values in the Economy* (Princeton, NJ: Princeton University Press, 2008).

33. Dan Chiappe, Adam Brown, Brian Dow, Jennifer Koontz, Marisela Rodriguez, and Kelly McCulloch, "Cheaters Are Looked At Longer and Remembered Better Than Cooperators in Social Exchange Situations," *Evolutionary Psychology* 2 (2004): 108–20.

34. Kevin McCabe, Daniel Houser, Lee Ryan, Vernon Smith, and Theodore Trouard, "A Functional Imaging Study of Cooperation in Two-Person Reciprocal Exchange," *Proceedings of the National Academy of Sciences* (98) (2001): 11832–35.

35. Joseph Henrich, Robert Boyd, Sam Bowles, Colin Camerer, Herbert Gintis, Richard McElreath, and Ernst Fehr, "In Search of *Homo Economicus*: Experiments in 15 Small-Scale Societies," *American Economic Review* 91(2) (2001): 73–79.

36. Joseph Henrich, Robert Boyd, Sam Bowles, Colin Camerer, Ernst Fehr, and Herbert Gintis, *Foundations of Human Sociality* (New York: Oxford University Press, 2004), 8.

37. Ibid., 49–50.

38. Herbert Gintis, Samuel Bowles, Robert Boyd, and Ernst Fehr, *Moral Sentiments and Material Interests* (Cambridge, MA: MIT Press, 2005); Robert Boyd and Peter J. Richerson, *The Origin and Evolution of Cultures* (New York: Oxford University Press, 2005).

10. The Science of Good Rules

1. First: Lon Haldeman: 9 days, 20 hours, 2 minutes. Second: John Howard: 10 days, 10 hours, 59 minutes. Third: Michael Shermer: 10 days, 19 hours, 54 minutes. Fourth: John Marino: 12 days, 7 hours, 37 minutes. The first three broke the previous transcontinental record of 10 days, 23 hours, 27 minutes held by Haldeman.

2. One more example reinforces the point. Since decorum soon breaks down along with the body and mind from extreme fatigue and sleep deprivation, cyclists tend to relieve themselves on the side of the road, sometimes even on the bike itself (for the guys, anyway). This led to an entire section in the rulebook on the matter of personal discretion, because some cyclists are more discreet than others. One year, for example, someone let it all hang out in front of a tinted restaurant window that obscured a clan of shocked diners (he thought the restaurant was closed). So we added a rule that if you cannot find a public restroom and your motorhome is not around, you can go on the side of the road only if your support crew surrounds you with an opaque sheet or blanket. This issue also led one of the top women competitors, who routinely led the entire race for the first thousand miles and typically finished in the top five overall, to calculate how much time she was losing to the men because they could pee off the bike whereas she had to stop and dismount. (She figured it was several minutes each stop, several stops a day, which in the course of ten days really does add up.) This led to the passing of a nondiscriminatory pee rule (later revoked after further consideration).

3. Thomas Hobbes, *Leviathan, or The Matter, Forme and Power of a Common Wealth Ecclesiasticall and Civil*, C. B. Macpherson, ed. (New York: Penguin, 1968), 76 (orig. pub. 1651).

4. David Landes, *The Wealth and Poverty of Nations* (New York: W. W. Norton, 1998).

5. Douglass C. North, *Structure and Change in Economic History* (New York: W.W. Norton, 1981); and *Understanding the Process of Institutional Change* (Princeton, NJ: Princeton University Press, 2005).

6. North's lecture is available at http://nobelprize.org/nobel_prizes/economics/laureates/1993/north-lecture.html. The economist Ronald Coase made the connection between institutions, transaction costs, and neoclassical theory. North writes: "The neoclassical result of efficient markets only obtains when it is costless to transact. Only under the conditions of costless bargaining will the actors reach the solution that maximizes aggregate income regardless of the institutional arrangements. When it is costly to transact, then institutions matter. And it is costly to transact. Wallis and North (1986) demonstrated in an empirical study that 45 percent of U.S. GNP was devoted to the transaction sector in 1970." See Ronald Coase, "The Problem of Social Cost," *Journal of Law and Economics* 3(1) (1960): 1–44; and John J. Wallis and Douglass C. North, "Measuring the Transaction Sector in the American Economy," in S. L. Engerman and R. E. Gallman, eds., *Long Term Factors in American Economic Growth* (Chicago: University of Chicago Press, 1986).

7. Erin O'Hara, "Trustworthiness and Contract," in Paul Zak, ed., *Moral Markets: The Critical Role of Values in the Economy* (Princeton, NJ: Princeton University Press, 2008). The link between contract law and markets has its early roots in medieval Europe. According to Indiana University political scientists David Schwab and Elinor Ostrom, the

medieval Law Merchant was "a set of legal codes governing commercial transactions and administered by private judges drawn from the commercial ranks. The purpose of these codes was to enforce contracts between merchants from different localities. A merchant who felt that he or she had been cheated by another merchant could file a grievance with the local private judge, who would then conduct a trial and, if the grievance was justified, enter a judgment on behalf of the aggrieved merchant." Much of the modern world is structured to deal with the very problems addressed by the medieval Law Merchant and to insure reliable transactions in the marketplace. See David Schwab and Elinor Ostrom, "The Vital Role of Norms and Rules in Maintaining Open Public and Private Economics," in Paul Zak, ed., *Moral Markets*.

8. Oliver E. Williamson, "The New Institutional Economics: Taking Stock, Looking Ahead," *Journal of Economic Literature* 38 (2000): 595–613.

9. T. Anderson and P. J. Hill, "The Evolution of Property Rights: A Study of the American West," *Journal of Law and Economics* 18 (1975): 163–79; John Umbeck, "The California Gold Rush: A Study of Emerging Property Rights," *Explorations in Economic History* 14 (1977): 197–226.

10. Robert C. Ellikson, *Order Without Law: How Neighbors Settle Disputes* (Cambridge, MA: Harvard University Press, 1991), 4.

11. Ronald Coase, "The Lighthouse in Economics," *Journal of Law and Economics* 17 (1974): 357–76. For a more general discussion of this and many other related problems, see Vernon Smith, "Constructivist and Ecological Rationality in Economics," available at http://nobelprize.org/nobel_prizes/economics/laureates/2002/smith-lecture.html.

12. Jared Diamond, *Guns, Germs, and Steel: The Fates of Human Societies* (New York: W. W. Norton, 1997), 286.

13. Ibid., 287.

14. *Inaugural Addresses of the Presidents of the United States* (Washington, D.C.: U.S. GPO, 1989); bartleby.com, 2001 (www.bartleby.com/124/).

11. Don't Be Evil

1. Philip Zimbardo, *The Lucifer Effect: Understanding How Good People Turn Evil* (New York: Random House, 2007).

2. Aleksandr Solzhenitsyn, *The Gulag Archipelago*, 3 vols. (New York: Harper & Row, 1973–78).

3. Charles W. Perdue, John F. Dovidio, Michael B. Gurtman, and Richard B. Tyler, "Us and Them: Social Categorization and the Process of Intergroup Bias," *Journal of Personality and Social Psychology* 59 (1990): 475–86.

4. Philip Zimbardo, "The Human Choice: Individuation, Reason, and Order Versus Deindividuation, Impulse, and Chaos," in W. J. Arnold and D. Levine, eds., *Nebraska Symposium on Motivation, 1969* (Lincoln: University of Nebraska Press, 1970). See also Zimbardo's *The Psychology of Attitude Change and Social Influence* (New York: McGraw-Hill, 1991). For a dissenting view of the connection between the Stanford Prison Experiment and Abu Ghraib, see William Saletan, "Situationist Ethics: The Stanford Prison Experiment Doesn't Explain Abu Ghraib," *Salon.com*, May 12, 2004.

5. Stanley Milgram, *Obedience to Authority: An Experimental View* (New York: Harper, 1969).

6. Solomon E. Asch, "Studies of Independence and Conformity: A Minority of One Against a Unanimous Majority," *Psychological Monographs* 70, no. 416 (1951). See also his "Opinions and Social Pressure," *Scientific American*, November 1955, 31–35.

7. Gregory Berns et al., "Neurobiological Correlates of Social Conformity and Independence During Mental Rotation," *Biological Psychiatry* 58 (August 1, 2005): 245–53.

8. Ibid.

9. Edmund Burke, *The Portable Edmund Burke*, Isaac Kramnick, ed. (New York: Penguin, 1999).

10. Interview with Phil Zimbardo conducted by Michael Shermer on March 26, 2007.

11. Joel Bakan, Mark Achbar, and Jennifer Abbott, *The Corporation: The Pathological Pursuit of Profit and Power*, documentary film, 2003. See also the book by the same title.

12. Michael James, "Is Greed Ever Good?" ABC News.com, August 22, 2002, available at http://abcnews.go.com/Business/story?id=85971.

13. Kurt Eichenwald, *Conspiracy of Fools* (New York: Broadway Books, 2005); Bethany McLean and Peter Elkind, *The Smartest Guys in the Room* (New York: Penguin, 2004). See also the documentary film based on the book under the title *Enron: The Smartest Guys in the Room*. For a skeptical perspective on the received view of Enron's demise, see Malcolm Gladwell, "Enron, Intelligence, and the Perils of Too Much Information," *The New Yorker*, January 8, 2007.

14. Clinton Wallace Free and Norman B. Macintosh, "Management Control Practice and Culture at Enron: The Untold Story," August 8, 2006, CAAA Annual Conference Paper, available at SSRN: http://ssrn.com/abstract=873636.

15. Quoted in Free and Macintosh, "Management Control Practice and Culture at Enron."

16. R. Bryce, *Pipe Dreams: Greed, Ego, and the Death of Enron* (New York: Public Affairs, 2002), 112.

17. Quoted in B. Gruley and R. Smith, "Anatomy of a Fall: Keys to Success Left Kenneth Lay Open to Disaster," *The Wall Street Journal*, April 26, 2002, A1, A5.

18. Robert Simons, *Levers of Control: How Managers Use Innovative Control Systems to Drive Strategic Renewal* (Cambridge, MA: Harvard Business School Press, 1995).

19. Alex Gibney, *Enron: The Smartest Guys in the Room*, documentary film, 2005, Independent Lens.

20. Ibid.

21. Quoted in L. Fox, *Enron: The Rise and Fall* (Hoboken, NJ: Wiley, 2003).

22. Quoted in P. Fusaro and R. Miller, *What Went Wrong at Enron: Everyone's Guide to the Largest Bankruptcy in U.S. History* (Hoboken, NJ: Wiley, 2003).

23. Quoted in L. Fox, *Enron: The Rise and Fall*.

24. T. Fowler, "The Pride and the Fall of Enron," *The Houston Chronicle*, October 20, 2002, 14.

25. No cats, please. "We have nothing against cats, per se, but we're a dog company, so as a general rule we feel cats visiting our campus would be fairly stressed out."

26. Robert Cialdini, *Influence: The New Psychology of Persuasion* (New York: William Morrow, 2006).

27. http://www.google.com/corporate/culture.html.

28. http://investor.google.com/conduct.html.

29. Although one corporate wag said evil is whatever Brin says it is.

30. Last updated January 30, 2007, http://investor.google.com/conduct.html#8.

12. Free to Choose

1. Read Montague, *Why Choose This Book?* (New York: Dutton, 2006). The title echoes Abbie Hoffman's classic 1960s social commentary *Steal This Book*.

2. Ibid., 2–3.

3. Ibid., 24.

4. Ibid., 16.

5. P. Read Montague and Gregory S. Berns, "Neural Economics and the Biological Substrates of Valuation," *Neuron* 36 (2002): 265–84.

6. Montague, *Why Choose This Book?* 99.

7. Friedrich Hayek, *The Fatal Conceit* (Chicago: University of Chicago Press, 1988), 68.

8. Samuel M. McClure, Jian Li, Damon Tomlin, Kim S. Cypert, Latané M. Montague, and P. Read Montague, "Neural Correlates of Behavioral Preference for Culturally Familiar Drinks," *Neuron* 44 (2004): 379–87.

9. Robert Lee Hotz, "Searching for the Why of Buy," *Los Angeles Times*, February 27, 2005.

10. An excellent discussion of the discovery of the D4DR gene and its implications for risk-taking behavior can be found in Matt Ridley, *Genome: The Autobiography of a Species in 23 Chapters* (New York: HarperCollins, 2001).

11. Benjamin Libet, "Unconscious Cerebral Initiative and the Role of Conscious Will in Voluntary Action," *Behavior and Brain Sciences* 8 (1985): 529–66.

12. Michael Shermer, *The Science of Good and Evil* (New York: Times Books, 2004), chap. 4.

13. Richard Layard, *Happiness* (New York: Penguin, 2005).

14. Ibid., 223–36.

15. Ludwig von Mises, *Human Action*, 3rd ed. (Chicago: Contemporary Books, 1966) (orig. pub. 1949), 860.

16. Interview of Mihaly Csikszentmihalyi conducted by the author on April 17, 2007.

17. Frédéric Bastiat, *Economic Sophisms* (Irvington-on-Hudson, NY: Foundation for Economic Education, 1996) (orig. pub. 1845).

18. http://www.brainyquote.com/quotes/quotes/g/georgebern128084.html.

19. http://quotations.about.com/od/moretypes/a/taxquotes1.htm.

20. Quoted in David Boas, *Libertarianism: A Primer* (New York: Free Press, 1998).

21. Albert Brooks, *Who Really Cares: The Surprising Truth About Compassionate Conservatism* (New York: Basic Books, 2006).

22. Ruut Veenhoven, "Quality-of-Life in Individualistic Society," *Social Indicators Research* 48 (1999): 157–86. See also his article "The Four Qualities of Life," *Journal of Happiness Studies* 1 (2000): 1–39.

23. Cass R. Sunstein and Richard H. Thaler, "Libertarian Paternalism Is Not an Oxymoron," *University of Chicago Law Review* 70(4) (2003): 1159–1202; Thaler and Sunstein, "Libertarian Paternalism," *American Economic Review* 93(2) (2003): 175–79.

24. Sunstein and Thaler, "Oxymoron," 1159.

25. Colin F. Camerer, George Loewenstein, and Matthew Rabin, eds., *Advances in Behavioral Economics* (Princeton, NJ: Princeton University Press, 2004).

26. http://www.prnewswire.com/mnr/fridays/27066/.

27. Sunstein and Thaler, "Oxymoron," 1191.

28. Barry Schwartz makes this point, as well as others, such as making MRI mammograms the default option that women must opt out of as the standard diagnostic tool. See Barry Schwartz, "Unnatural Selections," *New York Times*, April 12, 2007, A21.

Epilogue: To Open the World

1. www.futurefoundation.org.

2. By employing the hindsight bias, of course, it is easy after the fact to back into the historical record and find someone whose prescience seemed to presage the Soviet collapse or the rise of the World Wide Web, but for our purposes here, suffice it to say that both events took nearly everyone by surprise, and that's my point. For an apparent exception see John Mueller, *Retreat From Doomsday: The Obsolescence of Major War* (New York: Basic Books, 1989).

3. Robert Carneiro, "A Theory of the Origin of the State," *Science* 169(2947) (1970): 733–38.

4. Peter Bellwood, *First Farmers: The Origins of Agricultural Societies* (Blackwell Publishers, 2004), and Mark Nathan Cohen, *The Food Crisis in Prehistory: Overpopulation and the Origins of Agriculture* (New Haven: Yale University Press, 1977). The Neolithic revolution may have also led to a shift in parental preference for offspring quality over quantity, especially for those capable of mastering technologies that lead to an increase in survival. See O. Galor and O. Moav, "Natural Selection and the Origin of Economic Growth," *Quarterly Journal of Economics* 67(4) (2002): 1133–91.

5. Jared Diamond, *Guns, Germs, and Steel: The Fates of Human Societies* (New York: W. W. Norton, 1997). See also N. Roberts, *The Holocene* (Oxford: Basil Blackwell, 1989).

6. Robert Carneiro, "On The Relationship Between Size of Population and Complexity of Social Organization," in *Southwestern Journal of Anthropology* 23 (1967): 234–43.

7. Frédéric Bastiat, *Economic Sophisms* (Irvington-on-Hudson, N.Y.: Foundation for Economic Education, 1996) (orig. pub. 1845).

8. Napoleon Chagnon, *Yanomamö: The Fierce People* (New York: Harcourt Brace, 1992).

9. Ronald M. Berndt, "The Walmadjeri and Gugadja," in *Hunters and Gatherers Today*, M. G. Bicchieri, ed. (Prospect Heights, IL: Waveland Press, 1988).

10. Jared Diamond, "The Religious Story: A Review of *Darwin's Cathedral* by David Sloan Wilson," *The New York Review of Books*, November 7, 2002.

11. In subsequent ethnographic studies, other anthropologists found that these indigenous peoples also experienced a significant increase in Subjective Well-Being. In other words, they were happier because the physical and psychological burden of the constant endemic wars were lifted. See R. B. Edgerton, *Sick Societies: Challenging the Myth of Primitive Harmony* (New York: Free Press, 1992); and M. P. Ghiglieri, *The Dark Side of Man: Tracing the Origins of Male Violence* (Reading, MA: Perseus Books, 1999).

12. Robert L. Bettinger, *Hunter-Gatherers: Archaeological and Evolutionary Theory* New York: Plenum Press, 1991). For a literature review on intergroup violence and aggression, see Anne Campbell, "Aggression," and Roberg Kurzban and Steven Neuberg, "Managing Ingroup and Outgroup Relationships," in *The Handbook of Evolutionary Psychology*, David Buss, ed. (New York: Wiley, 2005); and Herbert Gintis, Samuel Bowles, Robert Boyd, and Ernst Fehr, *Moral Sentiments and Material Interests* (Cambridge, MA: MIT Press, 2005).

13. Rudolf J. Rummel, *Power Kills: Democracy as a Method of Nonviolence* (New Brunswick: Transaction, 1997). See also http://www.hawaii.edu/powerkills/pk.chapi.htm. For additional discussions on the relationship of democracy and war, see N. Beck and R. Tucker, "Democracy and Peace: General Law or Limited Phenomenon?" Annual Meeting of the Midwest Political Science Association; Steve Chan, "In Search of Democratic Peace: Problems and Promise," *Mershon International Studies Review* (47)

(1997); Christian Davenport and David A. Armstrong II, "Democracy and the Violation of Human Rights: A Statistical Analysis from 1976 to 1996," *American Journal of Political Science* 48(3) (2004); Michael W. Doyle, *Ways of War and Peace* (New York: W. W. Norton, 1997); Christopher F. Gelpi and Michael Griesdorf, "Winners or Losers? Democracies in International Crisis, 1918–94," *American Political Science Review* 95(3) (2001): 633–47; Hyung Min Kim and David L. Rousseau, "The Classical Liberals Were Half Right (or Half Wrong): New Tests of the 'Liberal Peace,' 1960–88," *Journal of Peace Research* 42(5) (2005): 523–43; David Leblang and Steve Chan, "Explaining Wars Fought by Established Democracies: Do Institutional Constraints Matter?" *Political Research Quarterly* 56 (2003): 385–400; Edward D. Mansfield and Jack Snyder, *Electing to Fight: Why Emerging Democracies Go to War* (Cambridge, MA: MIT Press, 2005); John M. Owen, "Give Democratic Peace a Chance? How Liberalism Produces Democratic Peace," *International Security* 19(2) (Autumn 1994): 87–125; and James Lee Ray, "Does Democracy Cause Peace?" *Annual Review of Political Science* 1 (1998): 27–46.

14. A good place to start is Irenaus Eibl-Eibesfeldt, *The Biology of Peace and War* (New York: Viking Press, 1979).

15. Michael Shermer, *The Science of Good and Evil* (New York: Times Books, 2003).

16. C. Gamble, *Timewalkers: The Prehistory of Global Colonization* (London: Alan Sutton, 1993), and T. D. Price and J. A. Brown, eds. *Prehistoric Hunter-Gatherers: The Emergence of Cultural Complexity* (Orlando, FL: Academic Press, 1985).

17. Shepard Krech, *The Ecological Indian: Myth and History* (New York: W. W. Norton, 1999), 152.

18. J. Rilling, D. A. Gutman, T. R. Zeh, G. Pagnoni, G. S. Berns, and C. D. Kilts, "A Neural Basis for Social Cooperation," *Neuron* 35 (July 18, 2002): 395–404.

19. This research was pioneered by Paul Zak at the Center for Neuroeconomics at Claremont Graduate University. See his Web page at http://www.llu.edu/llumc/neurosciences/. For the research on national trust, see P. J. Zak, "Trust," *Journal of Financial Transformation* (CAPCO Institute) 7 (2002): 18–24.

20. Thomas L. Friedman, *The Lexus and the Olive Tree* (New York: Anchor, 2000); and *The World Is Flat* (New York: Farrar, Straus and Giroux, 2005).

21. Michael Shermer, "Starbucks in the Forbidden City," *Scientific American*, July 2000, 34–35.

22. Robert Wright has a good discussion of this in his influential work on nonzero cooperative relationships in both history and life, *Nonzero: The Logic of Human Destiny* (New York: Vintage, 2002), 215–16.

23. Don Tapscott and Anthony D. Williams. *Wikinomics: How Mass Collaboration Changes Everything* (New York: Penguin/Portfolio, 2006).

24. A good summary of the growing globalization movement that was published as I finished this epilogue is Nayan Chanda's *Bound Together: How Traders, Preachers, Adventurers, and Warriors Shaped Globalization* (New Haven: Yale University Press, 2007).

25. Ludwig von Mises, *The Anti-capitalistic Mentality* (Indianapolis: Liberty Fund, 2006) (org. pub. 1956). Also available at http://www.mises.org/etexts/mises/anticap/section5.asp.

26. The quote has also been attributed to Thomas Paine, Abraham Lincoln, and others, including the Irish orator John Philpot Curran, who is quoted as saying, in 1790, "The condition upon which God hath given liberty to man is eternal vigilance."

27. The phrase served as the motto of the Institute for Human Progress, and I am grateful to Jay Stuart Snelson for bringing it to my attention.

ACKNOWLEDGMENTS

I would like to acknowledge the support of the Gruter Institute for Law and Behavioral Research, a private nonprofit institute that has fostered interdisciplinary research and teaching designed to inform law, economics, and other social sciences about the latest scientific findings about human behavior. Most notably I thank Monika Gruter Cheney and Oliver Goodenough at the Gruter Institute, and especially Paul Zak for sharing an early draft of his edited scholarly volume *Moral Markets: The Critical Role of Values in Free Enterprise*, also supported by the Gruter Institute and published by Princeton University Press. The John Templeton Foundation and the Ann and Gordon Getty Foundation supported of this work through the Gruter Institute.

Special thanks go to those economists, scholars, and scientists from various fields who took the time to read one or more chapters or the entire manuscript and offer constructive criticism and helpful comments, or to sit down with me for interviews and conversations, including (in alphabetical order): Colin Camerer, Randy Cassingham, Monika Gruter Cheney, David Cowan, Mihaly Csikszentmihalyi, Craig Fox, Oliver Goodenough, Stan Liebowitz, Stephen Margolis, Barnaby Marsh, Gerry

Ohrstrom, Russell Poldrack, David B. Schlosser, Jay Stuart Snelson, Paul Zak, and Philip Zimbardo.

As I have in all of my books, I would like to acknowledge a number of individuals who have contributed not only to this book but to my work in general, starting with my agents Katinka Matson and John Brockman, not only for their personal support but for what they have done to help shape the genre of science writing into a "third culture" of what I call integrative science, where one integrates data, theory, and narrative into a unified whole. And to my lecture agent, Scott Wolfman, for having the courage to market science and skepticism as a viable form of entertainment for college campuses, and for being such a good friend. Thanks as well to Paul Golob at Henry Holt/Times Books, who oversaw the project, and especially to the amazing Robin Dennis, my brilliant editor, who has greatly shaped my thinking and writing into a finer prose than I otherwise would have produced. I also acknowledge my remarkable copy editor, Emily DeHuff, who saved me much literary embarrassment with her many excellent suggestions, my production editor, Chris O'Connell, and the designer of the book, Victoria Hartman, whose typography, layout, and design elevated the book to elegance.

I also wish to recognize the office staff of the Skeptics Society and *Skeptic* magazine, including Pat Linse, Tanja Sterrmann, Stephanie Luu, Nicole McCullough, Ann Edwards, Daniel Loxton, Emrys Miller, William Bull; senior editor Frank Miele; senior scientists David Naiditch, Bernard Leikind, Liam McDaid, and Thomas McDonough; contributing editors Tim Callahan, Tom McIver, and Harry Ziel; editorial assistant Sara Meric; photographer David Patton and videographer Brad Davies for their visual record of the Skeptics' Caltech Science Lecture Series. I would also like to recognize *Skeptic* magazine's board members: Richard Abanes, David Alexander, the late Steve Allen, Arthur Benjamin, Roger Bingham, Napoleon Chagnon, K. C. Cole, Jared Diamond, Clayton J. Drees, Mark Edward, George Fischbeck, Greg Forbes, the late Stephen Jay Gould, John Gribbin, Steve Harris, William Jarvis, Lawrence Krauss, Gerald Larue, William McComas, John Mosley, Richard Olson, Donald Prothero, James Randi, Vincent Sarich, Eugenie Scott, Nancy Segal, Elie Shneour, Jay Stuart Snelson, Julia Sweeney, Frank Sulloway, Carol Tavris, and Stuart Vyse.

Thanks as well for the institutional support for the Skeptics Society at the California Institute of Technology goes to David Baltimore, Susan Davis, Chris Harcourt, Ram Basu, Debbie White, Hall Daily, Gail Wash, and Kip Thorne. Likewise, I appreciate the institutional support of the School of Politics and Economics at Claremont Graduate University, most notably Paul Zak, Thomas Willett, Thomas Borcherding, and Arthur Denzau. As always, I acknowledge my friends at KPCC 89.3 FM radio in Pasadena, most notably Larry Mantle, Ilsa Setziol, Jackie Oclaray, Julia Posie, and Linda Othenin-Girard, who have been good friends and valuable supporters of promoting science and critical thinking on the air. My friends Charles Bennett and Robert Zeps have been especially supportive of both the Skeptics Society as well as the skeptical movement in America, and I would like to acknowledge the generous support of the Skeptics Society from Jerome V. Broschart, Tom Glover, Matthew D. Madison and Sharon E. Madison, Ted A. Semon, Daniel Mendez, Robert and Mary Engman, and Whitney L. Ball. Finally, special thanks go to those who help at every level of our organization: Stephen Asma, Jaime Botero, Jason Bowes, Jean Paul Buquet, Adam Caldwell, Bonnie Callahan, Tim Callahan, Cliff Caplan, Randy Cassingham, Shoshana Cohen, John Coulter, Brad Davies, Janet Dreyer, Bob Friedhoffer, Michael Gilmore, Tyson Gilmore, Andrew Harter, Diane Knudtson, Joe Lee.

First among equals in friendship and support are John Rennie and Mariette DiChristina at *Scientific American* for providing skepticism a monthly voice that reaches so many people. I look forward each month to writing my column more than just about anything else I do in my working life.

Finally, I thank my daughter, Devin, for bringing so much joy just by being herself, and to my wife, Kim, with whom I have now shared life for two decades.

INDEX

ABOUT THE AUTHOR

MICHAEL SHERMER is the founding publisher of *Skeptic* magazine (www.skeptic.com) and the executive director of the Skeptics Society. A monthly columnist for *Scientific American* and an adjunct professor of economics at Claremont Graduate University, he is also the host of the Skeptics Distinguished Science Lecture Series at the California Institute of Technology (Caltech) and the cohost and producer of the thirteen-hour Family Channel television series *Exploring the Unknown*. About Shermer, the late Stephen Jay Gould wrote, "As head of one of America's leading skeptic organizations, and as a powerful activist and essayist in the service of this operational form of reason, [he] is an important figure in American public life."

Shermer is the author of numerous books, including a trilogy on belief: the bestselling *Why People Believe Weird Things,* on pseudoscience, superstitions, and other confusions of our time; *How We Believe: Science, Skepticism, and the Search for God*, on the origins of religion and why people believe in God; and *The Science of Good and Evil: Why People Cheat, Gossip, Share Care, and Follow the Golden Rule*, on the evolutionary origins of morality. He has also published two collections of essays: *Science Friction: Where the Known Meets the Unknown*, about how the mind works and how thinking goes wrong, and *The Borderlands of Science*, about the fuzzy land between science and pseudoscience. He is also the author of a biography, *In Darwin's Shadow*, about the life and science of the codiscoverer of natural selection, Alfred Russel Wallace, and *Denying History*, on Holocaust denial and other forms of pseudohistory.

Shermer earned his B.A. in psychology from Pepperdine University, his M.A. in experimental psychology from California State University at Fullerton, and his Ph.D. in the history of science from Claremont Graduate University. He has taught psychology, evolution, and the history of science at Occidental College; California State University, Los Angeles; and Glendale College.

He lives in Southern California.

Bestselling author and psychologist Michael Shermer
explores the science behind how we think
in these titles available from Holt Paperbacks

Why People Believe Weird Things

In this age of supposed scientific enlightenment, many people still believe in mind reading, past-life regression theory, and alien abduction. With a no-holds-barred assault on popular superstitions and prejudices, Michael Shermer debunks these nonsensical claims and explores the very human reasons people find otherwordly phenomena, conspiracy theories, and cults so appealing. As a champion of science and history, Shermer brilliantly exposes the all-too-human figures hiding behind the imposing curtains of myth and superstition.

How We Believe

In this illuminating study of God and religion, Michael Shermer offers fresh and often startling insights into age-old questions, including how and why humans put their faith in a higher power. Shermer explores the latest research of psychiatrists, neuroscientists, and epidemiologists, as well as the role of religion in our increasingly diverse modern world. Whether believers or nonbelievers, we are all driven by the need to understand the universe and our place in it. *How We Believe* is a brilliant scientific tour of this ancient and mysterious desire.

The Science of Good and Evil

A century and a half after Darwin first proposed a theory of "evolutionary ethics," science has begun to tackle the roots of morality. Just as evolutionary biologists study why we are hungry (to motivate us to eat) or why sex is enjoyable (to motivate us to procreate), they are now searching for the very nature of humanity. In *The Science of Good and Evil*, Michael Shermer explores how humans evolved from social primates to moral primates; how and why morality motivates the human animal; and how the foundation of moral principles can be built upon empirical evidence.

Science Friction

In fourteen essays touching on the barriers and biases that plague and propel science, Michael Shermer offers one of his most personal books. Whether he is pretending to be a psychic for a day or confronting impossible medical decisions for his mother, he displays the rewards of applying the scientific method to the challenges that face us and probes our ability to live both skeptical and meaningful lives. At turns thought-provoking, amusing, and heartrending, these essays consider the reach of science into the unknown.

Why Darwin Matters

Evolution is the most tested of science's great bodies of work, yet it is constantly under attack. Michael Shermer decodes the battery of facts supporting evolution and shows how natural selection achieves the elegant design we see in life. Shermer, who was once an evangelical Christian and creationist, refutes the pseudoscientific arguments underlying Intelligent Design and demonstrates why anyone—liberal or conservative, regardless of religious belief—can and should embrace evolution.

www.michaelshermer.com

www.henryholt.com